AF734403

THE CONSCIOUS NOTEBOOK

A NARRATIVE HUMAN ONTOLOGY

THE CONSCIOUS NOTEBOOK

A NARRATIVE HUMAN ONTOLOGY

RAKESH BISWAS

Nova Science Publishers, Inc.
New York

Copyright © 2009 by Nova Science Publishers, Inc.

All rights reserved. No part of this book may be reproduced, stored in a retrieval system or transmitted in any form or by any means: electronic, electrostatic, magnetic, tape, mechanical photocopying, recording or otherwise without the written permission of the Publisher.

For permission to use material from this book please contact us:
Telephone 631-231-7269; Fax 631-231-8175
Web Site: http://www.novapublishers.com

NOTICE TO THE READER

The Publisher has taken reasonable care in the preparation of this book, but makes no expressed or implied warranty of any kind and assumes no responsibility for any errors or omissions. No liability is assumed for incidental or consequential damages in connection with or arising out of information contained in this book. The Publisher shall not be liable for any special, consequential, or exemplary damages resulting, in whole or in part, from the readers' use of, or reliance upon, this material. Any parts of this book based on government reports are so indicated and copyright is claimed for those parts to the extent applicable to compilations of such works.

Independent verification should be sought for any data, advice or recommendations contained in this book. In addition, no responsibility is assumed by the publisher for any injury and/or damage to persons or property arising from any methods, products, instructions, ideas or otherwise contained in this publication.

This publication is designed to provide accurate and authoritative information with regard to the subject matter covered herein. It is sold with the clear understanding that the Publisher is not engaged in rendering legal or any other professional services. If legal or any other expert assistance is required, the services of a competent person should be sought. FROM A DECLARATION OF PARTICIPANTS JOINTLY ADOPTED BY A COMMITTEE OF THE AMERICAN BAR ASSOCIATION AND A COMMITTEE OF PUBLISHERS.

LIBRARY OF CONGRESS CATALOGING-IN-PUBLICATION DATA
Available upon request

ISBN 978-1-60692-735-9

Published by Nova Science Publishers, Inc. ✢ New York

CONTENTS

PREFACE

Present trend in computer ontology is toward the development of artificial intelligence and human ontology is an intervention to positively support it with natural human intelligence such that it is infused with the pluralism that characterizes human social structures.

This book plans to make a beginning toward that end utilizing a meta-narrative that examines the life of an academically failed physician and his relatively academically successful daughter. The short individual narrative beads that are threaded into this larger narrative represent a multi genre science and fantasy of medicine. On one hand it balances a postmodern stance with its incredulity toward absolute evidence based truth on the other a tolerant pluralism that simply recognizes all approaches as credible as long as the resultant is geared toward positive outcomes.

The book portrays a non-linear narrative ontology interspersed in linear discourses on its relevance to human cognition and ontology. It creates a fictional conceptual model of a human that is visualized in physical form analogous to a notebook computer that hides an infinite backend process of cognition analogous to the human mind.

This particular backend entity is labeled con (shortened from a consciousness that is universal to all humans and other sentient life also sometimes represented simply as being) and is fictionally endowed with powers that enable it to run on multiple notebook computers (analogous or metaphorically morphologically indistinguishable from humans). This helps the meta-narrative to explore human cognition and its physical manifestations in an Earthly plain.

The story line begins with Con taking the plunge with his notebook June into an Earthy material life. A plunge that makes him fall through life and is bound to end in death. Death smashes his notebook's hard disc into tiny bits and pieces. The nonlinear narrative as a result tries to pick up these broken fragments of memory and brings out Con's journey through the human body (in its tree like statistical self-similarity with the Earth and the universe, which may be represented in an atom). It portrays microcosmic interactions inside the human body at a macrocosmic level of day to day living on an Earthy scale.

Con becomes well versed with the anatomy and circuitry of the various intricate components of his machine but also realizes that it has developed in an evolving assembly line whose creators are long dead and nobody till date understands perfectly how the damn thing works. However there are theories, stories of atoms and molecules and their subatomic families regularly utilized to explain how semi conductor chips work inside our bodies. Throughout the ages, stories take on multicolored hues, theories on the nature of Con, Earth and self-evolving machines…the science is ever changing.

To all the dead bodies who taught us anatomy
To all our patients who taught us medicine
To all systems that need trouble shooting
Living or non living, sleek, shining, functional or worn, withered and junked

Sutra

We believe that the cognitive sciences have reached a situation in which they have been frozen into one narrow form by the machine metaphor. There is a need to thaw that form and move from a reductionist, atemporal, disembodied, static, rationalist, emotion and culture-free view, to fundamentally richer understandings that include the primacy of action, intention, emotion, culture, real-time con straints, real-world opportunities, and the peculiarities of living bodies.

Walter J. Free man and Rafael Núñez
Restoring to Cognition the Forgotten Primacy of Action, Intention and Emotion,
Imprint Academic 1999.

'You are not your body, but the witness, the consciousness of the body. Not any of your thoughts or feelings but the witness, the consciousness of your thoughts and feelings. '

--Line from, "The Mandukya Upanishad".

The Upanishads (translated by Swami Prabhananda 1970). Hollywood, Calif.: Vedanta Press. 1978.

INTRODUCTION

WHAT ON EARTH IS HUMAN ONTOLOGY?

The traditional goal of ontological inquiry is to divide the world "at its joints," to discover those fundamental categories or kinds that define the objects of the world. (IDEF5 methods report 1994). What ontology has in common with both computer science and philosophy is the representation of entities, ideas, and events, along with their properties and relations and the rules that govern them, according to a system of categories. ("Ontology-computer science", 2007) An often-quoted definition of ontology in computer science is that it is an explicit specification of a conceptualization restricted to different domains. (Gruber 1993).

For example if one has to write an ontology of driving a car it would involve representation and retrieval of data specifying certain concepts related to controlling the wheel, clutch, brake as well as concepts of objects such as road on which the car needs to be driven and characters such as trees, other cars and people that need to be avoided.

In human ontology, the ontologic domain is existence itself and for an individual sustaining that existence his/her personal ontology is his/her personal conceptualization (although rarely explicitly specified and this is where the similarity to computer ontology ends) of the roadmap he would need to drive himself through that existence and the objects s/he would meet on that road that s/he would need to know how to deal with. A computer ontology program for car driving may not actually drive a car unless connected to a robotic system but would likely be able to answer queries on car driving. In most humans substantial portion of the knowledge on car driving may however remain tacit where the body may know how much pressure to exert on the accelerator to attain a certain speed but for the individual may become difficult to tell. Instances where explicit specifications of conceptualizations are difficult to attain simulate real world human interactions where substantial outcomes are achieved in a tacit manner. (Biswas 2008)

Human ontology (a product of natural evolution) unlike computer ontology (artificially created by humans) is a remarkably heterogeneous cognitive phenomenon varying widely from human to human depending on genetic and environmental factors. It is a less explicit specification of individual concepts so much so that the term human ontology is a contradiction (unless it is just used to label neutral specifications for understanding human anatomy with an aim of building the healthcare framework around it). Admittedly the validity of claims labeling human creations as artificial is also debatable.

Yet humans do share cognitive constructs utilizing natural language that are expressed or stored as representations of entities, ideas, and events, along with their properties and relations and the rules that govern them, according to a system of categories (this is again what ontology is about although in humans it may not be explicitly stated as for computers).

Other contemporary views from authors such as Peter Wilberg deny human ontology as a cognitive process at all. Wilberg maintains that the inner human being cannot be reduced to a set of bodily processes or cognitive behaviours. The inner human being is not some bodily 'thing' at all but 'some-one': a being. According to Wilberg, human ontology creates a fundamental distinction between the human body and mind on the one hand, and the inner human being on the other. (Wilberg 2007)

Health as a personal experience is an individual phenomenon – patients understand and experience their health in the context of their social networks. People and events simultaneously make up these networks; structurally they are interconnected and functionally interrelated – and as such interdependent. (Bar-Yam 1997)

Over time individual patients therefore build their unique conceptual models of their health and illness in the context of their experiences and everyday social interactions. Equally health professionals develop their main constructs of health and disease from the collective interactions with their colleagues in medical institutions. Frequently these constructs are devoid of the 'white noise' generated by the personal realities of patients, caregivers and other professionals.

These personal individual views form the basis for the socio-cultural construct of health, illness and disease, and the culture of the health care system – these understandings are learnt, internalized and enacted. As Berger and Luckmann (1966) point out persons and groups interacting together in a social system, form over time, concepts or mental representations of each other's actions, and these concepts eventually become habituated into reciprocal roles played by the actors in relation to each other.

This book illustrates the dynamic interactions between individual human ontologies utilizing multiple beads of individual narratives threaded into a meta-narrative in the background of healthcare and illness.

REFERENCES

Bar-Yam, Y. *Dynamics of Complex Systems,* Addison-Wesley, 1997.

Berger, P. L. and T. Luckmann (1966). *The Social Construction of Reality: A Treatise in the Sociology of Knowledge*, Garden City, NY: Anchor Books.

Gruber, T. R. (1993). *A Translation Approach to Portable Ontology Specifications. Knowledge Acquisition*, 5(2), 199-220.

IDEF5 method report (1994) Retrieved Oct 7th 2007 from http://www.idef.com/pdf/Idef5.pdf.

Wilberg, P. Human Ontology or human genomics. Downloaded from http://heidegger.org.uk/hum_ont_hum_gen.pdf.

Ontology-computer science. (2007, Aug 12). In Wikipedia Retrieved oct 7th 2007 from: http://en.wikipedia.org/wiki/Ontology_%28computer_science%29.

Biswas, R, Martin, C.M., Sturmberg, J., Mukherji, K., Lee, E.W.H., Umakanth, S., and Kasthuri, A.S. (2008). Social cognitive ontology and user driven healthcare, Hatzipanagos, S. and Warburton, S. (2008). (Eds.), Social Software and Developing Community Ontologies, IGI Global Publishing (In Press).

THE CONSCIOUS NOTEBOOK: A COLLECTION OF NARRATIVES

All works of fiction draw on real life data and the present work is no exception. The data has been meticulously distorted substantially to remove all possible real life links. Any resemblance to a real life person would be purely coincidental.

Reading tips for the diligent but gullible reader

The trick in this book is not to be diligently linear as this is a very non-linear narrative built within the framework of a scale-free network with the sections on the medical student and her father as the major hubs. You could start with the last chapter and even end somewhere in the middle if you want to but be sure to turn it rapidly and smell the ink (that is if you are reading a paper version) or just eye the contents (not on the table of contents) scrolling down slowly till you feel like reading the italicized quotes (if you are on your electronic version).

Consider a social network in which nodes are people and links are acquaintance relationships between people. It is easy to see that people tend to form communities, i.e., small groups in which everyone knows everyone. In addition, the members of a community also have a few acquaintance relationships to people outside that community. Some people, however, are so related to other people (e.g., celebrities, politicians) that they are connected to a large number of communities. Those people may be considered the hubs responsible for the small world phenomenon.

http://en.wikipedia.org/wiki/Scale-free_network

PROLOGUE

WHO WANTS TO JOIN THE DANCE?

I don't know where the audience for this sort of a narrative collection or assortment lurks at the moment unless they are reading this but then why would anyone want to read it is the question that I must first try to answer. Also in my role of the prologue deity I guess I need to explain why they might consider sitting through the entire assortment of narratives.

I don't think I have to go further back in history than the cave men to search for this. Those were the good old days when primordial genes were discovering the joys of finding new expression in variety of human bodies that were beginning to work and think differently. Like one poor cave man would often be chided by his partner for not being able to pluck that bright red flower from the topmost branch of that tall tree, her friend's partner (called husband in the middle ages) having beat him to it.

So he would begin to think of obtaining the flower by offering to trade that neat piece of stone he has shaped into a sharp and a sort of business is born. This goes on for centuries until it's centralized into a governable currency and the modern market is born with the same underlying theme. Information as a commodity enters the market with the advent of the printing press and is presently threatening to leave with the advent of a world wide spidery market place.

I believe the biggest return for any information monger is information itself. It may in the beginning seem a bit like flattery (when someone conveys their gratitude for the information you presented) but eventually settles down to becoming an interesting exchange where both interacting poles gain more and more information.

Now, if one settles instead for getting rewarded in the form of a certain green kind of purchasing power then one simply manages to tire/burn oneself out. After all how much can you gorge with that puny body? In such situations the only hope of salvation is to open the floodgates of charity (albeit in a controlled manner) and achieve that equilibrium one strived and managed to lose in youth.

The real market lies in our hearts where, in the light of the full moon, one can dance to the tunes of a cowherd's flute, lilting the spirits of the jungle. The real buyer is one who can join the dance. In the end I must warn you, you are going to see more of the prologue deity barging in quite often in between the narrative.

Chapter 1

INTRODUCING DARKNESS

Your article says, "Something out there holds swarms of galaxies together and keeps their stars from falling apart, but scientists still haven't learnt what this invisible substance is." Although I am familiar with many of the names of God, this is the first time I have heard Him called Dark Matter.

Clyde A. Bachelor
Letter to National Geographic, September 2005

What you see in front of you is not just darkness. It's like you have no sight. Now imagine you aren't hearing me either. There's no sound except a soft drone, which drifts out from a deep black hole from time to time. Well… in such instances you might feel the soft rustle of air around or the saliva at the back of your tongue or become aware of the perfume your neighbor's put on but lets think that you aren't able to feel all this either…there's just space, no atmosphere to speak of, no stars coming out, no taste or touch, none of the five senses at all. Our story begins right here where time stands still…no sound of my talking, no pin dropping only a drone which sounds vaguely like aum…hmmmmmmmmmm.

Welcome to the black hole of Nirvana…universal consciousness. I know it's boring to be bereft of the senses especially when you know there's a thriving world full of revelers right in this black hole, who are like ordinary human beings talking, listening, jostling, tickling each other, enjoying their beers, having sex, in short having a real good time. Another thirty seconds of silence in this heart of darkness and you will understand the predicament of Con, the hero of our story. He's been holed up here since Con only knows what time…ages maybe but then again time here stands still…as if waiting.

Cons are actually ordinary people who had eaten the forbidden fruit on Earth. They wanted badly to merge into the main stream of universal consciousness, they also wanted peace and there was only one way out for them…unplug all sensory inputs from their system…with each and every bite on the apple, cut off all desires and reach the state in which all of you are now, if not for my incessant yapping. I shall stop here and leave you with nirvana for company just another 30 seconds…give you the feeling of what it's like to be Con (alias God)…poor God.

Tamas or Darkness in physical nature finds expression as gravity and inertial mass. In human nature it is felt essentially as a downward-pulling sense of inertia and heaviness. People search to compensate for Tamasic existence either through Rajas - hyperactivity and consumerism or mindless entertainment or through bland Sattvic states of spiritual harmony, peace and calm.

Peter Wilberg
The Gunas, 2007

Chapter 2

ONCE THE THREADS ARE SET...MUST CONTINUE TO WEAVE

All authors build on their own lives to tell their stories. They use their real life as an introduction to the characters and then sit back and watch their characters emerge with a life of their own. All this is done as part of this unique thought experiment called novel.

Sutra

A note from Sutra:

I am not to be mistaken for the author himself. The author to relate his story is using me. My Bengali name is Sutra Dhar, which doesn't necessarily mean storyteller but someone who threads and weaves a story. The things, which I relate, may happen in the real, the author's world and the imaginary, reel…my world. I am a bridge between the two worlds. The picture in front of you shows a man crossing the street. He's not the author or myself or even the sponsor for our story. He's an ordinary nameless human, the focus of our story.

He is an ordinary man who's weirdly fascinated by the ordinary people living around him in an extraordinary world. He revels in the stories churned out by their interactions with each other and the macro-microcosmic environment. He gorges on them daily as if they were a staple item in his daily diet of thought. I may not be the only storyteller on whom the author has bestowed sole right to voice his stories. There are others from the author's world of ordinary people. You might wonder whether they are people at all. You might get to hear stories related by a virus, a red blood corpuscle or even a protein molecule masquerading as a snake or squirrel. Sometimes you might hear the lord almighty talking to you.

Paradigms are fires lit by tiny sparks of individual thoughts that brighten up our worlds as a celebration and transform incendiary individuals into overnight life long celebrities. Society needs these glowing fires from its brightest stars to function effectively in a galactic space that is otherwise bereft of color and air in a back drop of intense all encompassing darkness—

Dhar S,
The influence of dominant paradigms in social functioning
(unpublished-retrieved from the dustbin folder).

Chapter 3

STORY OF SAMSARA

> To come back again to the beginning of time, even before the sperm comet hit our ovum planet, let's take a look at the source of our comet and planet. After all, the universe is just a collection of narrative matter strung together by time. –
>
> Sutra

Samsara, which means the universe, is only one of the characters in this story. He wasn't always bad in academics; in fact one should say he excelled in them. On leaving school, he managed to sail into the best technological institutes in the country, all because he had a way with mathematics. Numbers talked to him. They pulled at his hair, splashed water on him and teased him often sending him into frenzy…more so when he was presented with a problem, which was difficult and impossible for the average student.

It was off course rare to find him tied down to problems, which he always liked to call challenges, as he would have solved most of them in a jiffy. Unfortunately he also bagged at the same time a seat in medical school as his aggregate in physics and chemistry gave him an edge even if his biology was just passable. His parents decided that as his elder brother was an engineer he had to be a doctor.

His medical student days were spent in rigorous studies but recurrent failures. He wasn't actually studying the subjects' medical school demanded but almost everything else, which included lengthy tomes on theoretical physics, combinatorial chemistry and non-linear dynamics. He took a number of years more than the average medical graduate to pass his finals and barely managed to finish his internship. Most of his batch mates had long finished their residencies and fellowships becoming busy assistant professors of Cardiology, Gastroenterology and the like in prestigious universities. His school-mates way behind his mathematical abilities who had nevertheless got into engineering now held important positions in high tech academic institutes as well as the corporate sector.

His parents had by now realized their mistake and come to terms with the fact that their son wouldn't after all become a medical man even if he'd managed to scrape through medical school. They hoped he'd take up a government medical officer's job. His cousin was doing well earning hefty bribes signing false post mortem reports. He had already built a bungalow, bought a car.

However Samsara Sen. was made of different fiber. He believed in following his bliss. He had toyed with the idea of appearing again for the technological Institutes entrance exams during his med school days but somehow was then too absorbed with theoretical physics to even bother about a career in engineering. He believed his bliss lay in talking to his beloved numbers…Listening all day to the melodious chatter of mathematics.

> What drives us to learn? Is it our jest for reward in the form of feel good food for thought? We learn anything for these food for thought rewards and like animals in a circus seem to be constantly foraging for it.
>
> Sutra

Chapter 4

THE CLASS.... HARD CLINICAL ANALYSES OR A BIT OF VELVET, WARM TEA

Do you know what 'curriculum' really means?" Moe Zimmer berg '79 asks almost as soon as you meet him. In Latin it means "running in circles," he explains. Therefore, at the Tutorial School in Santa Fe, NM, which he founded with Richard Test in 1992, there is no curriculum. There is, however, plenty of learning—in almost any subject a student requests. Zimmerberg, who has an undergraduate degree in neuroscience from MIT, believes that people naturally want to learn, so classes are available but not required. "We offer structure, but not coercion," he explains.

An Alternative Education, Heather C. Liston,
Tech Rev, 04/02

Hi! I am still there...Sutra that's my name and what you see in front of you is a classroom. It's a class on trouble shooting humans the professor teaches in the University of Madeeasy. He's not the author mind you and don't keep asking me if that's his real world. Things might change or alternate so very rapidly in the story line I've been handed over to vocalize, even I might not be able to keep track of all the minor details.

I know I have a sort of upper hand in deciding which details are minor or major. That's the joy of the imaginary world or would you like to call it the virtual world? One is so very free to do as he pleases. In fact my free to do as you please tone might reflect the world I am frequenting at the moment.

Here in the visual before you, you see people, looking like real world students, politely and patiently hearing their professor's discourse. From time to time you can see them interrupting, questioning, themselves discussing at length on the topic. You can't hear them, nor can you see the professor. You can only see him through his students.

Ontology is also briefly described as a formal conceptual model. Each and every human forms his or her own informal personal cognitive conceptual model (informal human ontology) as a part of her/his social cognitive interactions.

Sutra

Chapter 5

SAMSARA'S BREAK- A NEW LIFE FORMAT

Finally he did get the break he was looking for right after he finished medical school, which had cost him more than a decade in his academic career. There was a Masters of technology curriculum that offered courses in biomedical engineering, which allowed medical graduates to join, provided they had basic knowledge of math and physics. He jumped at it and sailed through the entrance exams, the childish physics and math problems they had set for world-weary doctors who were expected to have forgotten most of their math. The very first day he was introduced to his guide who was an impressive asstt professor at both the local medical school as well as the technological institute. He had actually done his MD in medicine after completing a graduate degree in engineering. He asked Samsara to work on preparing a better variety of condom, not exactly a Herculean task if not for the fact that it required Samsara to measure the penile sizes of the average male in the area for whom the condom was being planned.

Samsara fell back on his quest for the truth in the non-linear chirping of numbers. He started chatting up the campus math professors who were reasonably intimidated by his stupendous mathematical knowledge, which by now had been flavored with creations of his own and lack of pedigree. He felt he should switch over to masters in mathematics but then being a medical graduate there was no way he could do that. In India people usually join medical school without a graduate bachelor's degree unlike the US. His MBBS degree wouldn't allow him to pursue masters in Mathematics.

An option in the institute open still was humanities, which allowed graduates to pursue masters irrespective of their undergraduate subjects. To escape the tedium of his condom thesis he switched over to masters in anthropology, which went well with his MBBS degree.

Imagine if people were numbers! That was the only way Samsara would be able to continue the anthropology course. If he could have his way he'd have converted all alphabets and written words into mathematical symbols. He sometimes wondered if people could communicate using only a universal language of mathematical symbols. He did his thesis explaining just this possibility and envisaged a hypothetical village where people talked in terms of mathematical interpretations of day-to-day life signals.

A lot of interesting papers followed once his thesis was published, with others citing his work as a primer to a future advanced mathematical civilization that for the time being started proliferating in journals if not on solid Earth. He became a prolific paper publisher and introduced novel mathematical concepts in anthropological data collection that nearly kicked out the traditional ethnographic approach. Alpha diversity measured as number of species and families of humans in a given plot along with their compositional change of species and families from one plot to another (beta diversity or turnover), calculated as a chord distance or dissimilarity index were old and inefficient for him. He longed for an index, which could measure and compare all the infinite data a person's day-to-day narrative generated.

Chapter 6

TO ALL MEMBERS OF THIS MAILING LIST

"Does the flap of a butterfly's wings in Brazil set off a tornado in Texas?" meteorologist Edward Lorenz once asked in postulating the "butterfly effect," (...). In today's world of the Internet the question might be rephrased: Can a single e-mail from Brazil set off a torrent of action in Texas?

Where To Start To Launch The 'Butterfly Effect',
Science Daily & Cornell University, 2004/02/16

An Email from "The Lord Almighty"

Dear List,

Call me Con for short. You can't see me and strictly speaking you aren't supposed to hear me either. I am speaking on a borrowed voice; this is an electronic voice mail.

You do hear my voice inside your head while you are reading this. You may be puzzled as this sounds like Sutra's voice as you have become familiar with it but I assure you Sutra's just a talking machine, a teller...don't confuse me with him. This is definitely Con speaking.

I just wanted to clarify a few things on this class of the professor. You might have heard about it from the teller and I am not a teller or professor myself but in plain and simple Conly terms, "trouble shooting humans", is an endeavor which is increasingly becoming very important for my day to day survival. Initially humans developed as a means of communication for Cons.

We were increasingly getting tired of the quantum ocean of pure consciousness with its irritating drone of Aummmmmmmm emerging from the depths like a burp from time to time serving as our only input. We decided we needed badly to have some fun. Bathe in a running stream for instance. Yeah! Yawn. The stream of consciousness beats the ocean any day...yawn. Gosh...those streaming videos last night were real tiring.

Social cognitive ontological constructs (SCOCs) are learnt over time since childhood under a variety of environmental influences that primarily shape his/her attitude. The SCOCs are also shaped by genetically inherited attributes such as learning aptitude etc. Most humans utilize their SCOCs in relating to the world around them.

Sutra

Chapter 7

Re: To All Members of this Mailing List

In this virtual world Con is the biggest online spammer.

Sutra

Aaargh! What a way to get back one's voice. Dear members of the list. Please don't believe a word that Con says. I've long since given up believing in him. Communications indeed! All He cares about is having fun with us poor human machines. Without us He doesn't have an existence at all.

He uses our five senses to the hilt whenever he can. All His idea of having fun! He never thinks of the consequences, the wear and tear we suffer. To think of the diseases He has wrought on us…we'll talk about the diseases in detail later. After all that is the purpose of this class…learning to trouble shoot humans. However before that we must understand the true nature of the Con human relationship.

Society also as a whole generates centralized social cognitive ontological constructs from collective social learning that trickle to the individual level in a top down manner.

Sutra

Chapter 8

CON HUMAN INTERFACE

Look at this slide...you might think you've seen this slide before. "An ordinary man crossing the street" slide. That's Con with his notebook, the walking and talking kind. Now you might ask me where his notebook is, as you obviously can't see one with him. Well actually he is the notebook. Con's own notebook...what you don't see is Con.

> Constructivist theories recognize that learning involves active and adaptive cognitive processing of individual experiences that may be even subjective and involves substantial social interaction. Cognitive processing is a cyclical activity of reflective observation on day-to-day concrete experiences followed by abstract conceptualization and active experimentation at either individual or group levels.
>
> Sutra

Chapter 9

CON'S BATTERY POWERED NOTEBOOK

As a "natural system" to organise biomedical data and knowledge, anatomy provides us with an order system that due to its physical, direct and "natural" appearance makes it inherently "user friendly". This central role of anatomy as the top level structural principle in biology and medicine is also highlighted by the fact that education of medical personal and medical doctors starts with basic courses in this discipline.

US National Science Foundation Strategic Research Workshop,
Digital Human, O n t o l o g y,
Bethesda, USA, 25-26 July 2002Workshop Report and Recommendations

The human notebook runs on batteries powering all its innumerable muscle cells generating outputs in all its activities of daily living. These batteries are again inside specialized cells organized into components concerned solely with channeling energy into the output generating muscle cells. Now you might wonder where Con gets all the energy.

The answer lies in the Sun, which supplies energy packets to trees. One of the activities of our notebook's daily living is devoted primarily to consuming energy through stuff that come from trees. All this stuff is fed into a pipeline with a wiring system of its own. The wiring delivers current that causes the muscle cells in the pipe to contract leading to an autonomic wave of peristaltic activity, which Con is oblivious of after his daily meals.

Apart from the notebook's auto-mode activities there's a generous allowance for Con's free will. Con hits the thought keys on his notebook for data input and it produces an output directed primarily at the muscle cells. Cells, which give us our facial expressions, propagate our chemical individuality.

Be it a smile or a dance maneuver, innumerable muscle cells of the microcosm sway like flexible bands attached to rigid skeletal rods on our macrocosmic notebook. They combine to produce a coordinated output manifesting as a cosmic dance. The last time Con remembers dancing like that was just before the lights went out. He was dancing like mad with his wife's corpse on his shoulders. Shiva dancing destruction, theorists termed it later and Taoist physicists compared it to the annihilation of the atom.

Chapter 10

AN EMAIL TO HIS STAR READERS FROM SUTRA

I am sending a copy of my manuscript a fictional post modern medical novel, the synopsis and a few sample chapters over to you by a separate mail. Could you go through them in your spare time and give me your valuable opinion, which would be required to modify it constructively/destructively even before I can think of a publisher. Do reply and let me know if you would like to see it following which I can send it over in a separate mail.

Thanks
Sutra

Dear Sutra,

The first feedback I got was unfortunately not positive. The very able physician-writer I asked for a review said despite his own interest in East meets West issues – the nature of your commentary, with religious overtones just didn't do it for him. The literary critic has given a preliminary response that was similar – again with some difficulty getting past the religious metaphor. He said he plans to read it more in depth and will give his final feedback then.

I'm sorry to pass along difficult news. Nonetheless, your concern for social justice shines through as far as I'm concerned.

Betelgeuse

Dear Betel,

I was surprised to know that my work appears religious. I thought I was the last person on Earth to be religious. Possibly it was my use of the Consciousness (alias God) Metaphor that was off putting and off-course my smatterings of Indian philosophy, which appears religious but is far from it.

Sutra

In general, most intuitive human knowledge still resists formalization, including that involved in comprehending simple stories or simple physical situations. Our surrounding environment has a much too complex structure that cannot be captured by a single ontology.

R˘azvan V. Florian,

Technical Report Coneural-03-01, February 4, 2003

Chapter 11

A COLLECTIVE CONSCIOUSNESS IN YOUR MAIL BOX

Hi, I am a human version of this guy named Sutra (in case you were wondering).

Let's forget about Samsara for some time and focus on his next generation. You can see this lady relaxing over a coffee spill resistant keyboard deep in asynchronous conversation with her friends.

So this girl June pecking at her key board is actually busy mailing her friends. This is still the era of the keyboard (I suspect you too are in the same era while you are reading this) and soon you may find humans freed from it particularly when you see them sitting opposite you in a train staring out into the receding background, in all likelihood they are sending and receiving mails through some kind of hidden computer inside their brains that respond to their thought keys.

What exactly happens when they compose and click on send is that it is being read by an artificially intelligent non human (our future cousins before they wipe us out altogether) and being further matched with similar email narratives with similar information content (and yet dissimilar in the sense that there is more to learn from the other information). Humans developed interesting software to do that starting with what was loosely termed web 2.0 and right now I have lost count which Web.digit it is.

This had enormous learning implications as you see June (or anyone else for that matter)

can share her lived experiences (mined from her daily emails) with anyone without being identified. This lets her communicate with a lot more individuals than if she were communicating with just a single person. She could receive daily feedback comments on her own experiences and thoughts contained in her mails. She could even choose to switch off this feature every time she sends a mail.

Even as you read this something of this sort is being developed in health care as chronic patient users start getting feedback from their health data contained in the lived experiences stored in their email boxes. As these users react to the feedback again by email as well as share their lived health records with their health professionals a new dimension of learning is generated in health care.

This user driven health care is helping health professionals like June understand their patients better as they go through the thematic analyses of lived experiential information in their patient records soon after they share their own health care giving experience on their day to day health informational transactions at the point of care again through the same email system which shows them matching experiences of other health professionals as well as patients.

You might wonder what health care has to do with collective consciousness...more on that again sometime.

Chapter 12

LETS BE NATURAL NUMBERS

Samsara got an appointment in the faculty of Anthropology in Sushunia University, West Bengal. His one successful paper had given him something of a celebrity status. He utilized it fully in having his own way with teaching as well, totally disregarding the set pattern of syllabi catering only to exams. It was probably related to his having failed a dozen times himself in medical school pointed out his colleagues. He wanted his subject to have a more profound relevance than just a means to an end it had become in his part of the world.

For most of his students the subject's only relevance was getting a professional degree, which would allow them to sit for the governmental competitive exams for the Indian administrative services. Samsara believed the key to a sustainable administration and human management lay in anthropological research, which would improve an understanding of humans to start with. Also he believed that human management was far from human control. That would require a shift in the attitude of present administrators and Samsara felt he could catalyze the shift by propagating his ideas through most of his students, future administrators.

"Let there be no control, he addressed a gathering of visibly alarmed and uncomfortable professors, no bindings of integrals or differentials, just let them be natural numbers and watch their natural history." His next flurry of papers was addressed to the issue of how not to help people actively and fall into the trap of having to control a cyclical bludgeoning debt that whirlpool you into a suffocating tunnel. It was mainly about how to simply observe and revel in the magic of the natural unperturbed oscillation. His favorite quote by Nicholas Shakespeare (which he noted down from a calendar) was, "There are too many people trying to change this world, who couldn't change a fuse."

His forays into human relationships led him into the adjoining villages around their Sushunia campus, a place with plenty of rock outcrops inviting a number of rock climbing groups from Calcutta especially in the winters.

One particular community in the villages that fascinated and fuelled his passion for the natural state was that of the 'Bauls'. They were singing philosophers who had the average human desires and yet radiated a flair for enjoying life through music. Somehow music was not immiscible with Samsara's mathematical passions. He had been trained in vocal Indian classical from childhood and had the ability to recognize the notes and classify them as soon as he heard a line from any raga. He didn't like the smoking and drinking habits of the bauls but when they sang he found the world he had been looking for.

Chapter 13

THE SERPENT BOSS

All life forms are composed of molecules that are not themselves alive. But in what ways do living and nonliving matter differ? How could a primitive life form arise from a collection of nonliving molecules? The transition from nonliving to living matter is usually raised in the context of the origin of life

Transitions From Nonliving To Living Matter, Steen Rasmussen, Liaohai Chen, David Deamer, David C. Krakauer, Norman H. Packard, Peter F. Stadler, and Mark A. Bedau,
Science Feb 13 2004: 963-965.

Here in this visual you see Con dancing on a serpent's head. Well it's not really Con but one of his symbolic representations human notebooks can ever hope to see. Looks like a black child holding a flute or is it a silent serpent whistle? The notebook is just an assortment of cells…cells that will turn to dust one-day…return to silicon mother Earth that is. Cellular functions, which give expression to Con's notebook, are due to one chemical individual, which resides inside a nucleus burrowed deep into cellular earth or cytoplasm if you fancy.

It's a huge coiled serpent, a fixed assortment of molecules, which populate the cellular environment. These molecules have been specifically combined permanently for each particular individual notebook. This serpent is what gives Con's notebook its chemical individuality. Deep inside the center of Earth, it peers out of its burrow and surveys the goings on like a ruling monarch keeping an eye on its subjects. Many molecules, responsive elements come to his nuclear pore window with begging bowls and he turns down no one.

At times he uncoils himself and holds out his own molecules to be copied into a complement RNA which like a faithful messenger goes out to the ribosome in the cytoplasm and works with another serpent RNA that transfers all amino acids near it into an orderly sequence as demanded by the code on its back. All these amino acids are bricks for building the serpent's empire that grows as the Earth grows. The serpent, Shiva's consort represents Shakti, the power that is necessary to drive all the energy Con's notebook acquires in its daily bread.

Yet genes alone can't specify the shape and arrangement of all an organism's parts, says Jacques Dumais, a biologist at Harvard University in Cambridge, Massachusetts. "There is not enough information in the DNA to code for that," Dumais says. "As soon as you talk about something more complicated than a virus, you can't do it."

Instead, the patterns of life must arise without detailed blueprints, just as an exquisitely symmetrical snowflake emerges from the random collisions of water molecules in moist air. A surprising resistance would be put up by sand grains hiding a buried treasure chest.

Life's Patterns: No Need to Spell It Out?, Adrian Cho,
Science Feb 6 2004: 782-783

Chapter 14

A ZILLION SNOOPINGS...HIDDEN FILES

The Samsara you see nowadays going on about well dressed with his wife is different from the Baulish vagabond he was before marriage. He had even brought a Baul to the classroom one day and made the students have a whale of a time while the baul fixed his cannabis on to the kolke (a pipe...sort of). He was lucky the principal hadn't barged in when the kolke passed on to the hands of some of his students. They had started taking polite swigs at the delicate cannabis that douses off if not puffed at frequently. They said they were just holding the kolke for the baul, while he sang a full-throated Lalan fakir. Here you can see some of Samsara's hidden files, which he has shared only with Maya his wife. I am sure he won't mind you all a peep.

Samsara's Diary June 3rd

Samsara they tell me is all *Maya* and that is what I like to think my wife is. We met for the first time amidst a few dolphins frolicking somewhere nears a remote island in the sea. We still have this custom of arranged marriages, not everyone is good enough to pick their own partners you see.

She looked too good to be true and I was a bit intimidated at first. Could she really like me? Then we got started talking and there was some magic working. The next thing you know, we decided to get engaged the very day after. Lobsters and other variety of fishes, there were so many its difficult to remember, entertained us with songs and food. The marriage date was fixed a bit too late for my liking. I had begun liking her so much that the next few months of wait were intolerable.

November 7th

We are in a train chugging through the sea. On both sides are a variety of trees bobbing up and down with fervor. There are more fields than trees. Fields of Arhar resembling a Bonsai jungle and wheat fields with a few sarsons thrown in between. Suddenly there's a gigantic wave seen from the widow with the trees too going up along with it as the train passes a mountain.

Chapter 15

THE NIGHT OF THE BED OF FLOWERS

November 16th

The last one-week has been hectic with the marriage and everything. Maya and me are one now sharing each other's existence and others too swimming nearby, exchange greetings. Maya doesn't talk much; she talks with her eyes that sometimes have a faraway look. I wonder what she keeps in those hidden recesses reflecting only the number of her hidden files.

November-? Undated

A white shroud over the groom…waiting…in anticipation…after the initial mantra chanting. Suddenly a blood curdling yell of ulu-lu-luuuuuuuuuuuuuuuuuu…! Generating images of African rituals in dense unexplored jungles… Through the shroud one sees a bright red bride being hauled in front of the pulpit. Her face covered in betel leaf, which she holds herself with an anxious diligence. She approaches seated on her magic carpet moving with remarkable speed. Her brothers holding the carpet (a pidhi in reality) were sweating under her weight. She was covered in a flowing red river with yellow sequins lighting lamps in diwali darkness. She was still desperately trying to cover her face; holding on to the pair of betel leafs in front of her, all the while trembling on her unstable perch. It was a game of hide and seeks and we were the chief players.

I felt sorry for her brothers who had by now started trembling under her weight. The game continued and fortunately they didn't drop her. It was a magic carpet after all. We were not supposed to see each other till that auspicious moment, when the pall bearers had wheeled her carpet around me for as much times as dictated. The rituals ended and I brought her home to my family. Till now we had only talked never loved. The first night after the marriage rituals is the night of darkness also labeled the night of time, which probably means where time stands still. It's a time when the boy and girl are not supposed to see each other and so there she was with the other members of my family, one end of the sari around her face, oblivious that I could see through if I looked hard enough.

The next day was the night of the bed of flowers. A bed decorated with roses of all shades, she sat pensive at one end of it…a bright red water lily blooming in a green pond. It felt strange to be in bed with a woman for the first time, a woman who's not my mother but may mother my child. I didn't know where to start and slowly started removing the red petals on the lily, watched them sail in frenzy through a raging mountain river finally set free from a stagnant pond.

Chapter 16

PECKING AT THOSE THOUGHT KEYS

Like Con himself was telling you…he just love's communicating, his ideas, his creativity etc. He wants to fill up our material world with his stuff, his doings…he wants to be called the creator of this universe. There are 33 crores of them littered around the place as old and dated Eastern mythology would have us believe. I have a feeling there's more of them than that and their number are growing everyday with their tendency to replicate assiduously producing swarms of thinking creatures who in turn love driving human machines.

If only Cons didn't have this peculiar habit of trying to communicate without seeing each other. Somehow seeing each other makes them repel. So they want to own human machines with a capability to transmit loads of information into a dynamic network of other Cons sitting with their machines. Initially Cons used the typing interface to input their information into the network. It was fairly acceptable getting a few mails each day from different parts of the world. However as human communication improved the typing keys were replaced with thought keys and a much faster new interface was born. Human machines started practically being driven by Con's thoughts.

> All our lives are spent in creating a cage of ontology specifications to enrich our day to day lives with the best possible outcomes and yet most of us deep down inside long for a window through which our inner bird flies out of this cage into a dream world. Some humans seem to have learnt the trick of moving in and out of this cage.
>
> Sutra
> Khanchar bhitor achin pakhi
> Kaemne aashe jai

Chapter 17

WHY TEACH AN OLD CON NEW...

> A giant universe snores... its chest inhaling and exhaling with time. Deep down in its lungs a few scientists write equations that arrive at this startling conclusion: the universe is expanding and contracting.
>
> Sutra

Humans have become so very indispensable to Con's thought processes in his virtual world that lately He has developed a nagging fear of what if...his priceless machine starts malfunctioning. Some of the thought keys haven't been working properly, thankfully not one of those alphabets he uses in his mail password.

Well, Con has realized the importance of his machine and at the same time feels insecure that nobody really knows how it functions. Given the limited life span of these machines He is increasingly aware that his communication prowess may be short lived. This is why he has decided to search out its maker but the other day he was quite shocked to find that it actually comes out of an assembly line where innumerable creators, nameless daily wage laborers sit together and assemble this marvel but nobody knows the whole. Con decides to find at least somebody who can repair it, somebody who has some idea as to how it functions as a whole.

> Most humans utilize their individually acquired (as a result of their learning behavior) social cognitive ontological constructs (SCOCs) in relating to the world around them. To visualize how our SCOCs look like just think of a stag with antlers. Now replace its face with that of a human and its horns with that of highly mobile antenna with infinite arborizations. Our individual SCOCs can make us attract, repel, laugh, cry and go through all the possible human emotions particularly when they overlap acutely with SCOCs of other humans (imagine stags locking horns) that although may have certain similar attributes, objects and relations in one plane or branch or node also tend to have very different attributes, objects and relations on another.
>
> Sutra

Chapter 18

FROM SUTRA'S READERS

Dear Sutra,

I read your manuscript towards the end of my holiday, and found myself bamboozled! I'm sure it wasn't the wine. The phrases seemed to come at me from every direction. I re-read the Introduction, and realized that you had successfully created a picture of the chaos into which Creation had degenerated and into which Con had to come to make sense of it and put it right. The message is great, but I think most readers would get lost and anxious in the chaos before finding their way out of it. Is there some way of writing the message about chaos without drawing the reader into it? Is there some way of using the theme to write about the chaos and anxiety felt by Con on engaging with the world, which would give a message of the humanity of Con to a reader, who could then identify with Con to whichever depth of emotion (and grief) they felt able?

Leo Alpha

Dear All,

I have one radical solution for the religious overtones. I have a feeling it's the use of the label God for our protagonist, which is most off putting. Most readers would have an image of God in their minds and it's this inbuilt image that is clashing with the image I am trying to project (that of human consciousness-something which most scientists and philosophers are trying hard to understand). So my solution is to call my protagonist by a different name. I feel substituting 'Con' for God would work well. I am sorry I couldn't recognize the religious overtones earlier possibly because I am so very irreligious.

Sutra

Dear Sutra

I look forward to reading this. Just so you know I'm actually very interested in reading your original manuscript in its entirety. Although I have not taken on the task of writing more than a few pages at any given time on any given subject ;) I've found that in verbal debate or 'spiritual' slanted discussion, I'm at times misunderstood as taking a religious stance, or I'm simply considered to be a religious person for bringing the word 'God' or the teachings of Buddha, or Theosophy (insert your favorite spiritual teaching/teacher here) to the table. Heavens knows (pun slightly intended) how many eyebrows I've seen raised when it's mentioned that I spent a month in Nepal at a Tibetan Buddhist monastery. The ironic part is that I've always considered religion of any sort to be somewhat of a hindrance to absolute spiritual growth. I've never considered myself to be religious (ok - I actually had a very strong belief in a white bearded Christian God until my early teens). I do however, have a strong interest and pursuit of spiritual knowledge, and religion of course is a rich, vast resource to tap into. So - - I guess all of that is to say, the word God doesn't 'scare' me, whether the word is capitalized or not. And I wonder if your target audience would feel the same way. Also, I'm just plain curious to read your manuscript as you intended it to be from your heart, pre-feedback).

Cheers, Rigel

Chapter 19

Pollinating for their Collective Bud

Samsara presumed he had done well the first night, the one with the bed of flowers. He disappeared immediately the next morning, which was a Sunday and roamed round with his Baul friends who visited various houses for impromptu sessions of chatting interspersed with songs, which he recorded in his walk-man. The first week was great for both Samsara and Maya who kept asking how long this would last. Samsara knew there was no point hurrying into her but then when Maya asked him suddenly one night, "Why don't you enter me?" he responded with an embarrassed silence and a hurried apology.

"You see, I haven't ever done it before and was taking time to get the hang of it". That seemed to satisfy Maya but Samsara brought it on himself to enter her immediately with his organ already limp after an external outpouring following the foreplay (this was becoming the routine in the last one week following their marriage). Samsara charged with his love making instrument and a determined look concealed by the darkness. It could only be halted by Maya's voice, "You are going too posterior, ouch!" It was far from the feeling, which had been lauded in books and Samsara thought of bettering himself the next night. He didn't retract his foreskin back to normal under the assumption that the receptors in his glans needed a bit of desensitizing for better charging and moved round the whole day with a definite uncomfortable sensation in his underwear.

The next night was a disaster. His glans far from being desensitized felt sore and entering became a painful ordeal but Samsara persevered with other discomfiting memories of the past in the background...his first smoke...uggh! The horrible taste and yet perseverance had allowed him to become a chimney, lauded and not looked down by his peers. This time his diligence didn't pay off and by morning he found his retracted prepuce swollen and paining like hell. He shuddered to think of the young cheeky surgeon eyeing both his penis and his wife while he listened to their story along with a host of medical students peering down at him and his battered ram. Fortunately Maya had a brain wave, got some ice from the freezer and stroked his preputial swelling into a manageably retractable size.

"What a relief!" after the prepuce finally covered his glans and his organ shrunk. After that he was wary of getting into the act for some days, entertaining the idea of an elective circumcision first but then one day it happened...the much-hyped feeling he was thirsting for. After a particularly intense night of vigorous foreplay when Maya became all limp his arms, he thought he'd give it another try and entered Maya with a little trepidation and a flaccid organ. It was a pleasant surprise to note that the more his glans stayed on her moist vagina the more it felt exciting with his organ going in a bit and coming out again slowly getting stronger and tougher. Mythical, tantric symbolic visualization of a glans in this context is more like an independent toddler playing in a temple courtyard, a temple that houses Shiva and his cobra with its hood licking the Linga, (glans) at each and every entry and exit, the toddler growing bigger, stiffer, harder and more and more exciting till it reaches its climacteric.

> Aaste aaste dhire dhire je kare gaman
> O taar hoi na re maran

Chapter 20

An Idiot Con's Guide to Humans

This is where I come in. I am the professor teaching this class of "Troubleshooting humans" in the University of Made-easy and I am not sure what Sutra's been telling you but it's been great going with the students since we began. Sutra's been collecting a lot of data along with the students for our story where Con's the protagonist with a lot of other animals and molecules that would appear to relate the narrative from time to time.

I have decided, I shall teach Con the works, all that he wanted to know about humans. I was debating whether to write a manual for them and call it an idiot Con's guide to humans, but decided against it. All said and done with all our knowledge of the macro and microcosm we still can't claim to know everything about humans. However we do troubleshoot them and perfect the art daily, tinkering with human machines in the hardware lab we call a teaching hospital. Our rounds start from the ICU at 8:00 AM and from today Con's invited.

Chapter 21

THE FAILED EDITOR IN CHEAP

Dear Sutra,

In your response you wrote "The concept of Con that I was trying to present in my book is quite different from the one that you have".

Would you please tell me about your concept of Con and what you believe spiritually?

You also wrote "My book which I had named, ' The Conscious notebook', tries to project Con as an ordinary human who has two parts…One - his consciousness that is Con and the other - his body that is his notebook." I have to admit...I didn't pick up on that in my reading and I'm quite confused on its meaning. Would you please share with me, how you decided on the title and explain, more in detail, what the above quote means...I'm lost on that.

As mentioned earlier, I'm a non-fiction reader. I enjoy reading medical reports on the most recent research, true life stories of hope and encouragement, and things on that order. My most prized possession and the thing I read most is my bible. Do you have a bible?

What is it like in your country? I'd really like to know more about your country and what you believe. It's a pleasure writing...isn't it interesting...we are on exact opposites of the world (I looked at the globe). We are directly opposite in the Northern Hemisphere, yet we are able to communicate this way. Grace and peace to you, Arcturus

Dear Arcturus,

Our concept about Con is likely to be very different but our goal is the same, this is to know Con and maybe become one with our creator. I follow an Eastern philosophy (but not very rigidly) that is very different from the Bible and is based on what are known as Upanishads, which are ancient texts that are as old as the bible if not older. These texts tell us that we humans are Con but we don't know it. Our one goal is to become one with Con. Once one is dissolved in an ocean of universal consciousness, where there are no boundaries and no concepts, no Bible or Upanishads only pure bliss and peace. We may be on different hemispheres but still live on the same Earth that is one tiny planet amongst the countless many in our universe with its innumerable galaxies. When you think of the cosmos in this manner you may get a little inkling of the enormity of being one with Con and losing all concepts of space time altogether. In the end, both of us realize Con can't be explained but may be felt and my story uses this Con=human consciousness model to satisfy the needs of the novel.

Sutra

Jeez Suts

I have so much fractionated reading to do of my own + learning Somali. I have failed you as editor-in-cheap. Ugh. Will do my best to read this; again, my inattention is just a product of life's overdose and fraying of neurocortices.......the Grouch (Capella)

Chapter 22

A NON PHARMACOLOGICAL BAUL TECHNIQUE

I hope persons with a rigid ontology don't review this book because if they do they may vomit on it and this would make that reviewer and others unable to read further.

Sutra

This was an idea Samsara got from Bhombol Khepa, a baul with an angta (hook) in his glans, someone who specialized in lifting heavy objects with it utilizing it to earn his living in fairs and festivals. "Don't worry about whether your organ becomes erect or not, just be calm. It's like crossing a bamboo pole across a raging torrent in a remote unexplored corner of the Himalayas. You've just got to steady your nerves…the prize lies in being allowed to continue your journey once you've been able to cross the first river.

A journey which takes you through deeper gorges, pristine valleys and higher snow bound passes until you reach the river again that needs to be crossed with you now wading deep into its spine chilling ice cold waters that feel like sharks biting into your shin bones (you almost pass out glad to be held back by the belay ropes)."

Bhombol continued, "After a time the pain subsides as your nerves become numb and you stumble into the opposite shore glad to have made it in one piece. The river is Saraswati, which flows through our spinal canal, a river that is no longer visible on the surface of the Earth, having long gone deep down the flower pot into its spinal tunnel sometime during Earth's embryo genesis.

All you can now experience of it on the surface is the breeze blowing through your nostrils, Ira Pingala, two sisters flowing through two of your external nares presumably coming and going from the spinal canal flowing as a river deep within your Earthy flower pot. It's the same river with multiple crossings, vital points in the journey from *Muladhar*(sacral center) to *Sahasradhar*(cerebral center). The penis need not be erect at the beginning of the journey. Just throw away the thought of consuming the pep up pill if it has entered your mind.

Relax and meditate on the river and the moon glimmering along it. It will awaken automatically once it's inside the vagina and you shall sway gently with the moon and the river making love in your boat. Samsara tried it on the very next day and made a mental note of his next paper, which would have the title, tentatively "Managing erectile dysfunction through non pharmacological ancient Baul techniques".

Deep down Samsara knew what Maya would think. "Don't think your mumbo jumbo is going to make you less of an old impotent fool that you are."

> Today we know of three important nervous centres responsible for the function of the penis; namely sacral erection centre, thoraco lumbar erection centre and cerebral erection centre. Fibers travel in the peri aqueductal grey mater to the cord where they connect to the thoracolumbar and sacral centres via the lateral columns. While many sensory inputs represent positive psychogenic stimulants, there are also many that are inhibitory.
>
> Weiss HD.
> The physiology of human penile erection.
> Ann Int Med. 1972; 76:793 9.

Chapter 23

FROM THE CLASS TO HARDWARE LAB... CHANGING DOMAINS

We are professionals and most of our waking lives inside the hospital will be spent on hands on teaching rather than lecture classes. Can't keep a busy outpatient waiting... Have to take a quick round through the intensive care unit and wards. Meanwhile you'll have to learn whatever you can by using your observational skills. There they are...my team of medical students, interns, house officers and consultants waiting to trail behind me. No point introducing you to them, they can't see you anyway.

Con started taking in the sight of various machines strewn around the ICU floor in reasonably comfortable beds, a chest pain, now drugged, A machine all puffed up and breathless, the case sheet diagnosis read--myocardial infarction day one, k four. A young lady producing a sound from her throat of the wind passing through bamboo reeds was appearing breathless due to the effort, a young man lying limp with a breathing machine attached to his throat seemingly pumping life. Con peered at the case sheet written by the night intern and got a shock. He had written a poem rather than anything substantial. It went, "Breathing machine pumping love, Eyes of dying hens, Pierce to depths of turbulence, Needles, blood gas and movement, Brisk sharp and painful, A cascade of alarms, ventilator settings, Causes and remedies, Eventually dropping off to an unusual silence, Mountain valleys, placid waters, Light of dawn and a distant bird song, Waking up to a humidifier alarm.

The professor continued, "I know to the uninitiated the ICU can be a ghastly sight with machines in a state of utter disrepair connected to funny tubings and monitors making awful sounds from time to time. Personally I feel you need to start at the bottom of the ladder here if you really want to know how to fix this stuff. You know, like becoming a medical student for instance. For that you will need a human body to interact with other fellow humans. You will need to acquire a fetus to start off with as a human baby, be part of a nice family while you grow up and then join this bunch of medical students (who might be your consultants by that time).

Unless you acquire a human form you can't really hope to be a good system troubleshooter. You'll just have to remain a silent invisible spectator from an indifferent Conly domain. Personally I am very much influenced by this saying," Life is something which everyone should try once."

Think over it while June tells you about the machines lying here. June is her Nepali name, which means the moon that can also suggest loony, and most of her descriptions will seem peculiar at first as you aren't used to the post modern medical jargon. She's the only other person in this group who can make out your presence here.

Chapter 24

HARDWARE INTENSIVE CARE

Con risked another peek at the case sheet written by the night duty intern for the patient on the ventilator. He had already gone through his postmodern medical progress notes and now examined his history sheet marked with a provisional diagnosis of acute inflammatory demyelinating polyneuropathy (AIDP). The history read, "One night a black raven swooped down, carried off all the power in his legs. Next day there were hordes of them, Screeching and perching at his hands and arms, a rude jab and he lost his voice, Another made him choke to swallow, He was brought to the ICU, Breathing slowly turning blue, A large ostrich perched near his head, Delivering mouth to mouth till he turned red, And in a discharge sheet he had made beforehand…he wrote, Gradually his power returned and he was out of the pumping station …power coming one by one in all his limbs. All birds one by one returned to their nests.

June was a lecturer and a good one at that no doubt, looking at the amount of awe and respect she commanded from the students and house officers alike although she looked like a medical student herself. She had thick glasses with dreamy eyes and wore her hair in lengthy plaited braids. She kept quiet while the professor discussed management strategies for the comatose patient in front of them. A few students raised their doubts based on the current clinical evidence to which the professor replied categorically, "The evidence always doesn't hold good while dealing with individual patients. In my experience I have seen this and my gut feeling tells me this and I have to apply this rule of thumb here. Unless the evidence base coincides with my gut feeling it's all rubbish."

While the herd trailed the professor as he moved on to the next patient, a wheezing lady, June stayed rooted to her place, bent down slowly over the comatose patient. She put her hand on her forehead and with her fingers flicked the patient's eyelids, which on opening, resembled a blank LCD screen).

The patient was a 32-year-old lady who had had a splitting headache followed by rapid loss of consciousness. June started rotating the patient's head from side to side all the while closely observing the patient's eyeballs move in the opposite direction (almost like that of a doll).

"The problem's in her cerebral cortices she whispered almost as if to Con. Her next few sentences reminded Con of the professor's warning.

> Our outer world external reality is ontology bound whereas our inner world surreal dreams are not. Humans who are more and more ontology bound may need to go often into

their inner worlds be it through meditation or sleep. Our inner worlds bring a sense of freedom from ontological bondage.

Sutra

Chapter 25

DOWN TO EARTH HUMAN NOTEBOOKS

June began, "All human machines in and outside this hardware lab are made of the Earth and all its elements (imagine your notebook 20 years from now as its silicon chips bite the dust…return to their elements that is). Most of their human lives are spent becoming mountains and rivers, which carry their life signals ("clearly beyond my domain,' thought Con). Look at this lady, as she lies oblivious of her surroundings, the river inside her is drying up fast. You can see it flowing very slowly and the monitor reflecting those slow waves. What was it that caused this problem?"

Con suddenly found June addressing him as if she was talking to a big class of students. "The driving force in solving hardware problems is asking the question why. Where is the problem? Once we know it's in the brain, we ask ourselves what's causing it? Could it be due to a disrupted interaction of all the chemicals in her brain, which utilize the rivers within to deliver goods to its various destinations? The river is what keeps her brain markets thriving, alive. Or could it have been a major Earthquake, a terrible landslide upsetting the course of her rivers, disrupting traffic, creating temporary lakes where they don't serve much purpose. We call them hematomas inside the brain and to us they are beautiful, at least they were so in the beginning when one was still unused to repeating patterns.

I still remember the first time I saw them as an undergrad student in my old hardware medical school back home in Kolkata. Off course one could only see them in pictures taken when the patient is wheeled into a giant camera which shoots pictures of the brain from all possible angles and then uses a computer to reconstruct them. We had a similar picture taken of this patient in our hardware-imaging lab downstairs and you can see all this blood, which looks so very white in between the grayish gyral regions of the brain and also in the ventricles that are natural reservoirs of cerebral fluid containing the same chemicals minus the blood cells which are restricted only to flowing inside the brain pipes. Now that the blood cells had burst into these reservoirs the pressure inside the brain had increased and this was causing her LCD screen to go blank as much as her neck and thigh muscles to go stiff.

Her CT head report looked like a newspaper clipping, "www.timesofindia.com,

Thursday, June 12, 2003, Fazilpur, Murshidabad, Badan Miya was a rich farmer with a large 2 story house. However nothing was left of it today after the raging river Padma had engulfed it. Over 1 million people affected by the river are waiting to be rehabilitated.

Chapter 26

The Drowning Differentials

Con let out a slow whistle. Wow! This trouble shooting business was cool. So this is how the brain goes wrong…hmmm! Wonder how it works normally…that would be interesting to know. "So every time the LCD goes blank does it mean blood in the cerebral fluid reservoirs?" Oh no! There are so many other causes…even this bleeding can be due to various types and sites of vessel rupture. There are rivers that have embankments, which may give away suddenly and unleash floods as a regular feature after the machine has lasted a good many years (sometimes I have seen young people die of it though). There's this terrible affliction of the river banks possibly the scourge of modern civilization. All those toxic wastes, which litter the walls of the river pipes along with the high pressure at which the river flows at present are good enough to burst any of them one of these days and then…bing goes the LCD screen.

Not only that, the litter and garbage we keep dumping into those rivers often cause these pipes to get clogged and you have the same effect depending on whether a major pipe is blocked. The pipes are a life line for the brain cells and with their food supplies cut off due to the river block they die and start rotting away, which makes the river toxic.

Then there are times when the meninges covering the brain cells catch fire. These are usually due to bugs, which create havoc with our machines from time to time. However very often it's the fight between the bugs and the immune cells which causes all the fire.

A person with meningitis has a burning body and a needle jab on his back to tap the cerebrospinal rivers would bring out a lot of inflammatory cells, a whole array of attacking soldiers. These immune cells are big guns but lousy shots…so much for their nuclear weapons. I sometimes feel they use too much firepower considering the size of their microbial targets but then it can't be helped. Whoever designed these machines was obviously not in his right mind. Going by their chaotic nature, there's an interesting talk about these having evolved spontaneously, like you know…people evolve into communities, villages and cities, expanding across the globe. Then there are occasions when the brain cells start dividing too often and produce too many more cells, all monsters at that, looking awful with their gaudy nuclear hyperchromatism and in general grotesque appearance. They lie accumulated together in one big tumorous community and can be easily spotted as an oddity on the CT head. Like fast growing cities eating into green Earth.

Then there are the tape worms that lay eggs which make their way into the brain for spending their larval infancy amongst the brain cells little realizing it's a dead end. They die and cause electrical short circuits in the brain cells and wham! The whole machine goes vibrating and shaking as if hit by a thunderbolt. It's strange that these worms should choose to survive only by seeking shelter in either pigs or humans (who are dead ends for them really) but then even humans couldn't have survived without having infected the body of Earth.

Chapter 27

CREATION: ASSEMBLING FROM JUNK

Embryonic stem cells can develop into many specialized cell types in culture dishes. It now seems that they can also generate primordial germ cells, which then go on to form sperm and eggs. A fertilized egg is potentially immortal: this fusion of egg and sperm gives rise not only to a new individual, but also (theoretically at least) to an endless series of generations. Three groups now suggest that it is possible to generate both of these remarkable cells known collectively as germ cells in a culture dish.

Stem Cells: How To Make Eggs and Sperm,
M. Azim Surani, Nature 427, 106 - 107 (08 January 2004)

The professor stepped into the pulpit and without a word to the 150 students started sketching on the board with a bunch of colored chalks. Their chirping had fallen silent at the suddenly noticed stealthy entry of this tramp, a silence punctuated by scratchy sounds of colored chalks on the black board. The professor of anatomy was thin built of average height and wore chappals apart from a large mass of flowing beard that clung like creeping orchids on his chin. Con thought they resembled a large stream of water falling from a high mountaintop, its droplets being swept away by strong winds in a bid to not let them touch the ground below (which was anyhow invisible).

His sketch looked somewhat like the moon with a crater dotted surface growing dense forests. The professor spoke for the first time addressing the air above his head (which was busy trying to transport his flowing beard), "What I've drawn is the uterus inside our body of Earth and what we are going to discuss is creation. At the moment we have a busy schedule of contracts pouring in to build notebooks by the dozen and so I shall have to keep my presentation brief. Right now we are busy preparing one for Mr. G (our latest client) who wants to become a medical student like you all.

In this wide expanse of the uterus within our Earth is a tubular place that opens into the uterus at the mouth of a narrow dark tunnel without lights except for a certain day of the month when its full moon. You can witness a busy market place being set up as it gradually moves from the tunnel into the lush green of the uterine valley with rivers and lakes (sinusoids) interspersed within.

Gharer madhye ghar banayia, Khelche sain ajab khela
Gharer madhye milaiche, Chander mela

Its moon market all the time with Con or so he wishes it would be. He just can't live without inputs. Look all round him and you will find him surrounded by dreams. He doesn't discriminate them with reality. To him anything which gleams like moon market is fit to be in his itinerary. There from his vantage point of the moon he can see the dark space being invaded by a beautiful comet with a tail like that of a sperm and then before he can say "what the…" the comets landed into his vantage point which is at present all rubble and chaos. Peering through the dust one sees things going as usual at moon market. If at all the transactions seem a bit more vigorous.

Takhan bayu kone jhar uthile, Dil dariyai uthbe dheu.

Chapter 28

FIRST RAIN LAST NIGHT; POO-WEE, YUCK, DRAT...END OF AN ILLUSION OF ETERNAL SPRING

At this point in the narrative there was a pause as the professor started wiping his glasses and then looked at his watch. He was newly married and his wife would be feeling hungry sitting in front of a loaded table waiting for him. He couldn't make her wait any longer. The story of embryogenesis could wait. As far as he could tell there were still wide gaps in their knowledge of human geopolitical evolution, the Earth and all its life forms over the ages. The evolution of their married life was different. It was a journey on a bicycle punctuated by frequent jolts due to minor conflicts lying strewn around like pebble-stones on the road.

The first time he detected a peculiar expression bordering between anger and anguish in Maya's eyes was when they moved into their new faculty quarters a few days from their marriage. Samsara had to join duty, his other beloved and so the whole job of setting up the furniture and various knick-knacks of their belongings into order fell upon Maya. Later at lunch, Samsara marveled at the neat homely look and said his friends wanted to come over but then there were no sweets in the freezer. After all they were a newly wed couple and guests had to be sweetened to the root of their teeth as the custom demanded. Personally Samsara detested the idea. Sweets didn't do any good to anybody. Sweets entering one's system left an overworked pancreas going broke on its insulin more so if its genetic program controlling the replication of beta cells producing insulin was destined to switch off after a time. Sweets were obtained nevertheless after a quick shopping and Maya agreed the guests be called over. Samsara didn't notice the dark clouds developing near the horizon.

All the while the guests were there Samsara noticed with a growing anxiety Maya's face which seemed like it was cut off from the party, forlorn withdrawn, like a gloomy cloudy evening thirsting for rains. There were tears in her eyes for the first time after the guests left and this was the beginning of a series of episodes coming cyclically every month ending in tears like a perennial monsoon.

Marriage is a blending of two separate cognitive entities that may not share the same sex but often share similar attributes that are enough to put them in the same car toward completing life's journey. Most partners once inside the car are often confused as to if they made a mistake getting in because they start imagining as if they have irreconcilable differences in their choice of destination. After all how can you continue in the same car if you want to go different places? Most humans don't realize that there is only one natural destination to life's journey and that is death. It is when we get too fond of the individual attributes in our carefully constructed life ontology we confuse these ontologic elements as our life destination conveniently forgetting that these are just short term destinations/goals, strategies to delude ourselves from the overwhelming uncertainties of the entire journey of life and death. Having said that it is understandable that most of us at times may think we would be happier leading our lives on our own ontology and may choose to act on those irreconcilably disparate attributes to go ahead with a painful separation.

Sutra

Chapter 29

ANATOMY- HARDWARE JUNK LAB

Anatomy lectures usually begin with embryogenesis. As Con stepped into the class, he found quite an assortment of 150 students chirping excitedly. Probably it was more due to the fact that this was going to be the first class of their undergrad medical curriculum.

Con wasn't born yet, that is he didn't possess a notebook (the walking and talking kind) of his own at the moment. He was unable to communicate in a human manner with his fellow students. As far as he could tell he was invisible as long as his thought processes could not be communicated.

He needed a hard disc to store his thoughts, a good CPU supported by a competent RAM to activate them into a communication either through his email or voice mail out-boxes. As long as he couldn't do that his thoughts would remain as diffuse Brownian particles traversing infinite distances in the universe bouncing back to and fro, something which other Cons with notebooks in their present state of limited functionality (the students) wouldn't be able to perceive.

So Con remained invisible and the constant chattering of the other students reminded him of his moon market. He walked out of the classroom and into the hardware lab, which was labeled "Dissection room" on a worn out wooden board. The professor was peering into a freshly dismantled notebook with all its electronic wizardry dancing out in a splendor of artistic circuitry.

Con watched open mouthed guzzling in as much of the visual delight, looking forward to the day he'd know his way in and out of all those circuitous pathways that resembled dirt tracks and rivers interspersed in a lot of green and brown. Before all that he would have to have a notebook of his own, he'd have to be born. He decided to check out his thought waves in moon market on the latest status of his notebook, how far they had finished working on it.

Moon market or Chander haath is a marketplace for exchange of ideas in return for expression of interest (EOI).

Sutra

Chapter 30

SPRING, FINALLY, OVERNIGHT BLOOM!!!!!!!.

If you go back far enough, humans, frogs, bacteria and slime moulds share a common ancestor. But scientists can't agree what it was like, or even whether it was a single creature.

Origins Of Life: Born In A Watery Commune, John Whitfield,
Nature 427, 674 - 676, 04/02/19

E' genesis

The ovum and the pronuclei...a husband and wife settle down into a moon within the Earth that is going to be their home for a few more generations to come. The husband, who has just landed in a space ship, loses its acro-plasm and greets the female pro-nucleus. They exchange words, an infinite number of genes that contain all the information they would like to have shared between them. A whole civilization grows as they roll down through the tunnel into a beautiful uterine valley replete with dense forests where they make a clearing. The family has already grown into a 16 member strong blastomere most of who share the same likes and dislikes along with their common dwelling. Their house sprouts roots into the uterine soil and Con watches dumb founded as his machine starts happening right in front of his eyes. All he can make out is a lot of cells being generated spontaneously as if growing from the other and moving off to different corners of their house. All this happened as if those cells were people with minds of their own. It seemed kind of cozy in there with all those people huddled together by the fire sharing stories. Gradually their stories grew as their numbers increased and their exchanges became more and more complex. People started to move, gone were the days one could simply till his patch of land and relax over a smoke. The world was being globalized and the new world belonged to those who were specialized, who had focused into developing particular functions among themselves. There were neural cells that would mainly restrict themselves to information technology. There were cardiac cells that were busy building one of the most ambitious pumping supply lines ever conceived complete with a network of rivers zig zagging across mountains. These rivers would carry all imaginable consumables for daily living. This central pumping system even had its own intranet so that it could keep beating even if severed of all connections from the neural Internet.

Tell you what...we are nothing but an assortment of countless bacteria thriving in their specialized roles (muscle cells, brain cells, cardiac cells...you name it) inside this home we call the body. Those bacteria who haven't found a role in our body are regularly passed out in our stools (would you believe it they form a major part of the fecal matter you pass daily). Some people claim the same to be true of the Earth with its remarkable colonies of free living entities however most would agree that in this Earthy scheme humans have proved to be the most pathogenic for Earth.

Sutra

Chapter 31

WHATEVER YOU WISH IS WHAT YOU GET

Con moved around inspecting his new machine that was going to be his home for some time to come on the Earthy dimension (he couldn't tell how much time now…here time stood still). He was off course mollified by the fact that all these cells, all these walking and talking qualified DNA holders were working to build HIS notebook, his home and he was the captain of the ship.

There was one hitch however. None of his workers seemed to recognize him. They appeared as if they couldn't care less about Con. They lead their own lives in their specialties, organ systems of heart, lung, kidneys, liver, brain etc and kept working like assembly line workers (that is as much as they were allowed by their DNAs). It was as if their lives and deaths were programmed into the blue print in their nuclei, that omnipresent serpent boss coiled in an upward spiral leading to an unending vortex.

Their lives seemed quite ordinary and Con couldn't figure out how these people would establish that vital link between all the organs which would make his machine finally take off as a whole, like a gigantic ship pushed off the sands into a deep blue.

He noticed there were some representatives the people comprising these organ systems had elected, like dutiful citizens of sovereign republics and these representatives were ready to discuss their terms with Con. In fact they were quite a confident bunch and sometimes their authoritative tones would even override Cons. Thankfully most of the time all they did was asked for more, more supplies, more rivers.

All said and done Con really didn't have to do much for them. It was a self organizing system, growing day by day, the cells initially coming out from their stems knowing exactly where they wanted to go in life, what they wanted to specialize in. Gradually the initial flurry of activity died down and most of them became trees, sessile and packaged near each other into dense forests, tissues they called them. In fact the trees were not the whole cells. They were just a projection from their bodies like receptors and most of these cells were like Earth. The kind of Earth you dig your hands and shape into anything you could wish for.

Chapter 32

NEEDED URGENT: PASSIONATE LOVER FOR A BORED WIFE

Maya's tearful voice was still ringing in his ears, "What I need is a lover." She was lying slumped on the bed in naked darkness and Samsara looked dazed (if only one could have a glimpse of his face through the dark that is).

The outburst had come because he had ejaculated before entering her and Maya suspected foul play. Samsara knew he was guilty of having avenged her reckless jab at him when she had recently declared, "I have never enjoyed sex with you. I feel terribly used at the way you paw me. You are not sincere and passionate in your actions…Samsara knew he had fallen asleep on a few occasions while stroking her hair or scratching her back (in what was supposed to have been a forerunner to a grand entry) but then again he had held himself well inside her on countless occasions and yet never seen a glimmer of satisfaction in her eyes ever (he wouldn't have minded even if she just feigned it). "…Therefore I had decided to let you enter only on my fertile periods. I need this child…otherwise life with you is becoming unbearable."

"Well if she feels used at my pawing why should I let her use me", he thought. That night as his hands played frenzy on Maya's body he took time to enter, actually he wasn't planning to at all but then he hadn't planned ejaculating either. The bed was a mess but that wasn't the only reason that he couldn't sleep.

Throughout the night Maya lay slumped naked emitting crooning sounds…Samsara assumed she was with the imaginary lover from her childhood fantasies. She had told him once of this man's face that was vivid in her mind and how her heart would flutter if she noticed anyone faintly resembling him…also how Samsara didn't fit his type.

Samsara brooded the whole day…there were a few options before him…get a legal separation, mutate himself into the man of her dreams or just find the man of her dreams and do a threesome. He rejected all these and came up with a fourth one that was close to the third. From today he would try to make friends with her imaginary lover. Maya could use Samsara all she wanted. He would even encourage her to close her eyes and visualize the face of her lover while he worked at her.

Commonly emotions of grief or relief are also generated when an individual acutely disconnects or reconnects with whom s/he has been sharing social cognitive ontological constructs (SCOCs). Also in the same individual when there is an acute shift in attributes contained within his/her own SCOCs, it may lead to a sudden outburst of humor, grief depending on the social context. A sudden coming together of two humans of opposite sex or containing SCOCs with different physical attributes generate sexual passion or infatuation depending on which physical attributes share focus although in such instances the interaction may be tacit as well as cognitive.

Sutra

Chapter 33

FALLING INTO A DIFFERENT DREAM

And then finally one day the machine was ready. At least Con could feel a signal coming from all their individual components merging into his own. Slowly he felt he could gather the collective consciousness of all these cells that had become the Earth with its trees and rivers. May I have the first slide? (Sutra)

You can see it's a sunny day there with a blue sky forming the backdrop of a giant redwood forest and Con finds himself on the topmost branch of the tallest tree…a tree so tall that it appears to have disappeared into the sky. May I have the next slide please?

A huge swarm of Lilliputian ants pushing Con's body over the precipice, which resembles the bark of a tree and one, can see him in free fall. Next…

At first in slow motion one sees the ground below approaching closer and then suddenly in real time there's a surge of panic as one hits rock bottom and everything becomes dark. The fall signals the beginning of his journey. It's going to be some time before Con really hits rock bottom although he may be reminded of it from time to time. Till then he was falling into a different dream…life they called it.

Surprisingly the landing wasn't that bad. There was hardly any impact and Con felt he had landed on a waterbed soft and bouncy. He started jumping up and down in glee. Maya sat up in bed and signaled at Samsara, "Shhh, listen…put your hand here and dragged his hands onto the bulge in her abdomen. "It's quickening, coming to life."

Impact or no impact he retained absolutely no recollection of where he had come from. He had even forgotten about being the captain of his ship with all the numerous cells he was commanding. Nevertheless everything worked with clock like precision as he kept on having the time of his life, jumping up and down in his waterbed.

He felt this was his true calling, he felt great…alive at last…having all the fun he had been thirsting for. It was still dark inside the womb but nothing like the awful darkness he had experienced

Earlier (all those years in the ocean of universal consciousness where there was no way to tell time as it stood still, waiting)…humph! Nirvana indeed!

> All this talk about social cognitive ontological constructs don't you think is a mouthful too tad to even begin to chew? It could have been expressed in a simpler manner by just stating that human emotions are generated in different states of being (philosophically ontology maybe simply expressed as a theory of being).
>
> Sutra

Chapter 34

SWISS CHEESE HOME DELIVERED

Professor Samsara was happy with the newcomer inside his wife. He had (or was it a she?) definitely brought some stability to the bad weather their marriage seemed to be running into. Once when the newcomer had stopped moving Maya was up the whole night and early next morning both rushed to their obstetrician who did an immediate ultrasound, which showed a heart beating perfectly (oops! What a relief). There were rounded things below the abdomen that made the obstetrician casually remark, "Looks like a male child". Samsara presumed he was probably hinting at the testes. In the last few months Maya bloated up like a balloon on her mother's cooking. Her mother was staying with them for a few months of the time of delivery.

Samsara prevented his own weight gain by taking longer walks. Maya's obstetrician said the baby looked small for her age (inspite of all that weight gain!!) but reassured them, "Small babies are easier to deliver". Samsara read up all the latest on fetal growth restriction on the net and from what he could gather it may have been a result of some blockage in the food supplying rivers from the mother to the fetus which circulate through the placenta, a condition termed placental insufficiency. The literature was sparse on why this happened in some mothers. It was causing all the goodies which Maya's mother made to fill up only her own fat depots giving her that unsightly erratic bulge. The ultrasound showed the placenta all riddled with holes like a Swiss cheese. Samsara made a mental note of examining her placenta once it was delivered. Most animals swallow their own placenta right after the birth of their children and Samsara wondered aloud jokingly to Maya if she would like hers well grilled or rare.

Finally D-day arrived when Maya went into labor. She seemed to be in terrible pain but bore it rather well clutching at Samsara's arms for comfort. Samsara thought he would give her a smile of encouragement but it became more of a laugh and certainly didn't go down well with Maya's tears at that very moment. Con barged in onto the second stage of labor with a force that cut it short by a substantial number of minutes. There was a vaginal tear in Maya as a result of which the obstetrician had to spend quite some time giving finer stitches. Samsara thought it must have been Con's razor sharp nose. Con! She's beautiful... He forgot the fact that she was supposed to have been a boy. Months later his obstetrician friend confessed he had erred banking too much on the testes that were actually intestinal loops... a phallus would have confirmed a male if he had seen it and there were other criteria of assessing for females. He also forgot about the placenta that probably lay in a corner of the nursing home dustbin, unexamined Swiss cheese, ungrilled, uneaten.

Chapter 35

NEONATAL ANTICS...LCD DREAMS

> It's one of nature's great insults that she should prefer to put all her eggs in the basket of a defense less, incompetent neonate rather than the in the tried and tested custody of our own superb minds. But as our neurofibrils begin to tangle and that neonate walks to a wisdom that eludes us, we are forced to give nature credit for her daring idea.
>
> Oxford handbook of clinical specialties

"What does Con see with his eyebrows contorted in his sleep, smiling even before he learns the social smile...is it like, he sees what he always wanted to? Things that were promised, that made him accept this defenseless puny body as his own?"

Having said this, the professor straightened himself on the ledge. He could barely manage a foothold on the precipice but he had himself anchored safely. He was also being belayed by one of his students most of who were climbing behind them looking like a host of geckos on a wall from a distance. "Whatever it is he sees it must be far better than those Nirvana days I am sure, better than complete darkness or a complete white out. All these years he had thirsted for a single color out of its myriad violet, indigo, blue green, yellow, red spectrum but no you have to accept the emptiness he was told for that was the absolute truth."

The professor resumed climbing again delivering an angry kick with his crampon front points biting into the ice as he hauled himself up with his ice axe. "All our mortal lives we try to cloak this emptiness, this void with colors and stories, weaving blankets covering our neonatal exterior. Blankets pile up one on top of the other while we wriggle our puny bodies kicking our feet in glee. We can't live without these stories. We have to be constantly surrounded by them, their warmth, like a pleasant whir of the electric warmer churning vibrations that in turn generate patterns of sounds we interpret as pleasurable. Most of us have to put in a certain effort to maintain our story lines. We have to feed our characters, take all their major decisions and remain in a constant fear of wearing out our stories prematurely exposing our emptiness in black or white. The very thought that there can be a world bereft of color is killing in itself.

They had arrived at a technically difficult point in their climb, a tricky traverse over a narrow ledge overlooking a fall of a few thousand feet. The professor set about trying to fix ropes using as many natural anchors as possible but also reluctantly hammering in a few pitons. He resumed his yapping with the students trying to listen attentively amidst fear howling like the wind roaring at the ground somewhere around the drop below.

"We can't be without work for long. Work is what gives us our daily narrative, the vital blanket against winter chills. We operate from the security of our blankets, our accomplishments, woven in narratives of our past promising warmer blankets for the future. The neonate must learn to read the stories reflected on his LCD screen in all its dazzling riveting complexity. His very survival depends on it. They are just a preparation for weaving a net of bigger dreams. Their absence would only remind him of the blank void, the fearsome darkness or irritating whiteness that awaits him at the very core of his being."

Chapter 36

GAMING...DAILY KICKING THE INVISIBLE FOOTBALL AMIDST US

"Yes, we were talking about poor Con who's terribly afraid of this emptiness, this silence at the core of his being and is desperately trying to immerse himself in a toy which has evolved over the ages and that is the human body (his notebook with limbs). Trouble is, once he's given the body to play with he starts thinking he is the body and forgets completely that he's different. He loses himself so very thoroughly into net surfing in the web and answering emails (He revels in his friend's narratives) that he becomes aware of his body only when it malfunctions. He's petrified if he doesn't receive emails for a day or if his computer keys simply refuse to type in what he wants it to. He realizes the need to know more about this machine and trouble shoot it himself if necessary."

The professor watched all his students traverse the ledge safely taking care to see if the karabiners were screwed tight. He unfixed the ropes in the end as he followed them at last into the relief of a wider shelf where they'd pitch tents for the night. "However its a tedious job, the prof began as all of them snuggled into their sleeping bags with their hot coffee...knowing about machines, no one seems to know much about them but more people seem to assume they know a lot and that's even more dangerous. It's a game no doubt as much as it's a story perhaps...as long as you can believe that I am supposed to be narrating it. There's one chief player who is one and at the same time many, who's playing with his machines one at a time (spontaneously at the same time) and if it's a game of football then the ball is at times invisible. Its like you give an input and it goes through a series of weights and synapses, all individual players kicking that input signal, their only goal being an output which does surface eventually but then the ball goes back again into the field starting from the input (by a back propagation algorithm) and it goes on. One gets tired of playing the same game day in and day out and so desires a change every now and then. The input output and back propagation signal remain the same only the football changes from soccer to rugby, even volleyball and the ground shifts to office working hours and cocktail parties.

Right now you see me...I am playing a game with my boss who happens to be a top notch consultant physician and we are going round fixing human machines strewn around the ward in a state of disrepair. Both of us are lunging at the ball like its some kind of latest bit of evidence which might hold the wonder cure for this broken down machine running on artificial hormones, secretions synthesized (nowadays in petridishes...genetic engineering marvels). More often than not the evidence would just be a bit of a thing about how this new drug's going to make a teeny weenie difference (if you add it in numbers it might look impressive and respectable) or how its not going to work...to think that most of our days go in undoing all the techno junk laid daily on us...All the same each and every game that we play tells us a story. They are the lifesavers playing on the background of our LCD screens, our LCD dreams.

> How does an individual write/form his/her own ontology for maximizing his lifetime rewards?
>
> Is it just a trial error process of experiential learning through social interaction that begins in infancy?
>
> Sutra

Chapter 37

A Mound of Honey Tipped Neurocortices

Everybody was back in the classroom looking fresh from the shower they had all had after a game of volleyball. The professor drew a square on the blackboard and rubbed it off.

"Is it not possible to do without this screen saver, this need to know the story of whatever and whoever we spot around us?" Well, sex may be one area where this is possible perhaps! Off course not that kind of sex, which is heavily dependant on a story line for arousal, like you imagine you're sleeping with someone else's wife while all you're doing is sleeping with your own wife like every other day. No it should be the kind of sex where you aren't really aroused by your thoughts (the usual source of your ubiquitous screen saver churning stories by the dozen).

It would be that state where you are simply engrossed in the mechanical act of copulation, concentrating only on the sensations being generated as a result of all that mechanical friction. It might lie in that state when you are all encompassed by that sensation, that pure bliss, which is unfortunately unsustainable for more than a few minutes before it's drained away in a spasmodic orgasm. Just because it can't be held for long we try to seek refuge in repeating it whenever we can... our only hope of escaping the dreadful narratives on our screen saving lives. There are other moments perhaps...like have you ever looked into the eyes of someone beautiful and simply kept on looking, taking in just for the pure sensory pleasure of it, not being plagued by thoughts of wanting to know who it is...what's the story behind this person?"

With that the professor bent down behind his desk and pulled out a swaraj (a small sarod) and started playing it, the notes on the strings dancing out into the afternoon sun basking inside the classroom.

> Aage na jene mojo na pirite
> Jene shune korbe pirit
> shesh bhalo darai jaate

"I don't know, the very objective on one hand in being a human seemed to be to dwell in the pleasures of the five senses but somewhere down the line we forgot how to do just that. Down the years, our senses were overridden by our capacity to think and create stories. Stories…erupting like molten lava creating a mound of honey tipped neuro-cortices over our thalami.

O kaacha haari te o haari te
rakhi te narili premojal.

Chapter 38

A LOT TO WEAVE, FIGURE OUT, JUGGLE

"You might ask me what I have against stories and their narrators; after all I started off as a narrator myself. It's the shameful media hype they enjoy, which gives them a power that taunts my powerlessness. Just imagine! Writing a story that captures a fan following and their imaginary screen savers, the narrator becomes so very powerful, so much so that just because he titillates our cerebral cortices, we allow him or her to control our minds. However all good stories must come to an end and then it's again back to the feeling you get an hour past your orgasm (unless you were tired enough to have dozed off). This is the biggest set back to our storied lives, as our thought processes keep swaying under control of our invisible innumerable narrators doling out a daily dose of inebriating story lines.

All scientific papers that have had some success have always depended on a strong narrator who could create a great story with a mass appeal. Even mathematical figures are known mainly by their association and interactions with each other, generating stories in all their riveting complexity. To escape all that and their antecedent problems, chief among which is the transient nature of the bliss it provides, we need to abolish all these stories being generated continuously on our consciousness screens. We have to abolish any thought, which arises as a curiosity about most objects we encounter on our sensory input panels. Just look at that person without thinking who he might be. Savor that mountain or waterfall without wondering about its geopolitical evolution in time. It's difficult...really after all we have already evolved from a thalamic, pure sensory perceptive level to a cortical level where stories abound...ready to capture us in their complex nets.

Real life is quite different from cortical life however. It consists of fragmented descriptions falling short of becoming stories. That work…is usually completed by our cortices. So it's the interplay of real life and cortical interpretations that gives us a reel life, captured in our hard drive memories all in B mode real time."

"Let's see if it's possible to describe real life without falling into the story trap. Real life descriptions necessarily come from past events and as because they are stored in our hard drives; they have already been compressed into a story format. Stories are essential for our memories. They are the only way we can zip those files, it's more like weaving or tying them tightly with our cortical ropes intertwining them with all the complexity one can generate. I shall begin with my original unfinished story and consciously try not to end it (zip it with a story line that is)."

The professor wiped his glasses after removing his eyes from the telescope. They had all gathered at the rooftop of his house for stargazing. They were all members of the sky watchers association of Sushunia, the students and the professor being the founder members. A lot of other people young and old joined them on their rooftop on fixed days of the week, the gathering becoming more voluminous on days when newer comets would be visualized, particularly if they were splashed into the public imagination through front-page newspaper photos. He resumed his talk again in the darkness below an all encompassing black umbrella studded with glittering sequins. "We have to reach a compromise. Stories are an essential cortical activity and they shall keep emerging on our cortices from time to time. The other part, delving in appreciating pure feeling as it is, rests in sub cortical processes, which may lie anywhere in the thalamus or limbic system or even somewhere down the spinal yogic chakras. Everyday we need to switch on and off between our cortical stories and sub cortical experiences and whatever emerges is life."

Chapter 39

ANATOMY AND PHYSIOLOGY OF RELATIONSHIPS

"Each and everyday as professionals our work starts with romance as we meet new people, identify new patterns and as the day progresses we get married to our work and get into a love hate relationship with it, which is not as exciting as romance but can be absorbing nevertheless until we are reminded of lunch. Then through the afternoon and evening begins a pure sub cortical activity of immersing oneself in that feeling which comes from a repeated mechanical friction on our sensory nerve endings till we reach orgasm by night and it's all over for some time when we sleep, that is when we have reached our refractory periods.

Every story begins in a romance, like a torrent gaining momentum after having freshly melted from the glacier. By midday it reaches a valley and widens out into a river bed rubbing against a lot of rocks sleeping on it managing to transform them into smooth stones resembling glans like shiv lings."

The professor paused for breath as the group paused to admire the view of the river snaking along the valley. Con was strapped to Samsara in a bag, which he carried in his chest while on his shoulders, rested a heavy rucksack carrying clothes for Maya and the infant. They were on their way to Annapurna base camp. The students had been looking forward to this trip since a long time.

"Our life begins in romance and necessarily has to end in marriage (that stands for any steady relationship breeding familiarity). It lacks the thrill of romance and can be trying in its own way. Nothing can fire us better than romance. Love is blander in comparison. It simply grows on you, takes time and is hardly ever noticed unless you are separated." Samsara paused to let Maya take over Con and put her to the breast. "This cycle of romance, marriage and orgasm is intricately built into most aspects of human relationships. Our first meeting with most people culminates in romance (provided it gets a positive feedback). The romance stems from the freshness of the relationship but gradually we become familiar and get married into a love hate relationship. Eventually we get preoccupied with the daily sub cortical mechanical to and fro sensory stimulation (sex for short) that occupies a major chunk of our social and private lives. More often it usually just ends with an orgasm and the much loathed refractory period leaving little choice except sleep."

Everybody was relaxed after the last part of the climb, which had been particularly difficult. They were sitting and sipping on to their Thumba at a cafe on the top of the mountain. This was the best part of the climb Samsara thought drinking in the view of Macchapucchare at the same time sucking at the bamboo reed dipped into his jar of Thumba.

"It all begins with cortical romance, descends down into a sub cortical orgy ending in a spinal reflex orgasm. Some people do claim its possible to hold this spinal energy and redirect it back to the cortex going through a return journey via all the spinal stations from Muladhar to Sahasradhar, spinal sacro coccygeal segment to the honey tipped neurocortices, its one hell of a Salmon like journey upstream on the river Saraswati flowing through the spinal central canal. All of us play this game of short-circuiting cortico spinal neural circuits in human relationships. A game we were destined to play from birth to death, between two goal posts in this huge stadium of life. When we are not playing, we become the audience munching pop corn in our refractory periods. "

That night they had bamboo shoots for dinner under moonlit Macchapucchare gleaming bright.

Chapter 40

THE DECISION TREE

"All humans aspire to be trees one day, perhaps because they are trees inside. Each and every one of them carries cells arranged in a tree like pattern. Look at the spinal cord, how it blossoms into the skin and muscles in our body from the roots of our brain, or the bronchial tree blooming alveoli forming two full lungs or the arteries, veins or lymphatics, the list of our trees within seems endless. Then there is the decision tree of life where we follow our birds hopping from branch to branch hoping we can keep pace with the growth of that tree."

The professor was showing them around lakeside introducing them to a lot of birds, which collected there in the winter. There were plenty of colorful waterfowls swimming and on the trees in the jungle near the lake were even more colorful characters in all shapes and sizes. Con could hear a particularly interesting call like that of a person hitting on a piece of metal. "Ah! That's the copper smith", remarked the professor and excitedly pointed towards a patch of crimson yellow. Con could see a small bird just bigger than a sparrow with peculiar red-rimmed specs around its eyes that somehow reminded him of June in the ICU.

Con began following the bird from one branch to the other until he lost it but found the professor waiting for him at one branch where he was sure he had last spotted the bird. The professor resumed, "I am a wanderer, a non-linear input signal in a network, static in a dungeon, yet marching on an un-trodden path, through jungles, rice fields and obscure Mountain passes. Each day I trudge along in the hope of a better life, some visual delight or the other, a newfound titillation or rediscovered magic. I have not ceased to travel, in spite of heavy odds. Rest yes, for one or two days whenever I felt like it, but my faithful dream was to listen to that melodious stream. The joy within us, which comes in all colors and black and white packets, sometimes they have a package insert describing their mellifluous qualities, sometimes none at all."

Con looked at him bewildered at this sudden poetic contemplation of the professor with the students also nowhere in site. "Young and old alike carry this stream all their lives, yet scarcely do they ever pause to listen, the sounds of a falling leaf, the raptures of a flight of birds and the gentle descent of a gliding night. Many a times have I despaired at the wanderer in me, is it purely a quest born out of hunger and thirst, something physiological underlying the emotional desire for more. I pause to look back and marvel at the immense complexity, the branching patterns and arborizations my chosen path unfolds, like the giant oak tree, which greets me from one of the windows of our library...Makes me think of the wanderer that was within its long forgotten seed."

Chapter 41

AN OLE CRONE, GRIZZLED AND RAWBONED

Con followed around the professor as he hopped from one branch to the other all the while spouting forth a torrent of free flowing obscure prose.

"What desire…or hunger was it born of…to reach its present magnificence; the brilliance of its architectural splendor arose from what internal yearnings. As its branches arborize chaotically towards an ocean blue sky, it portrays a full-grown image of a resolute tree, determined to live a life well earned, chaotic and yet, simple and elegant as it strikes the eye.

However old age descends with time, age that withers, shatters all complex patterns, all the hard work of youth so painstakingly carved on each offshoot of every branch. So with disease that wreaks havoc with the body's intricate branching. All paths traced out on sand, all castles swept off by that surreptitiously incoming tide. All our lives, born out of earth turn to dust… with a single deft brush stroke. Yet there is something, which remains in the dust, which gathers on earth, with each passing day, as the sky turns crimson at dusk only to herald a pinkish dawn. There comes a time when the complete spectrum of colors splash over with renewed vigor on an often used earthy canvas and paint anew another masterpiece.

Most of our lives are spent in becoming trees and mountains, carving our own masterpieces selfishly guarded. Only part of it is spent in undoing them. Each of us longs for an exquisitely beautiful piece of work that we can call our very own built from earth. Yet how cruelly do we ignore our alternative possibilities…things we could have done. The same time if spent in some other work would have led us to other branches. I wouldn't have lamented on your lack of love for me. That would have been just another branch of a tree amidst other innumerable branches. Countless possibilities in which you and I would have lived happily ever after or would have never met…all branches…separated only by time. All my life I hopped from branch to branch and scarcely did ever notice the tree in its entirety…a spectacle that would dazzle us.

> A stochastic process, or sometimes random process, is the counterpart to a deterministic process in probability theory. Instead of dealing with only one possible 'reality' of how the process might evolve under time, in a stochastic or random process there is some indeterminacy in its future evolution described by probability distributions. This means that even if the initial condition (or starting point) is known, there are many possibilities the process might go to, but some paths are more probable and others less.
>
> http://en.wikipedia.org/wiki/Stochastic_processes

Chapter 42

LETTING YOU INTO OUR WHIRLWIND

Maybe trees too think the same about us, our numerous arteries branching into capillaries meeting venules…rivers of life. A white blood corpuscle within us doesn't even know how intimately it's related to a distant neuron bound by same flesh and blood. We are so engrossed in our local surroundings, so prone to disregard non-locality with impunity.

We forget, our lives, which arise and perish in a continuous cycle of rebirths, are but one life…one only and the rest are simply patterns. There can arise from this pattern nothing essentially new. Evolution is just a rearrangement of what all ready exists in the future perhaps, but even the future exists in time. It's the river of time, which allows or forces us to distinguish these individual patterns.

What then are breakthroughs? Are they simply discovering a new branch? There have been many schizoids who've walked into new branches…its not enough to discover new branches, one has to have a fan following. New nests have to be built on newer branches, not alone but with many others. The more people one influences better the breakthrough. One must remember however, all our discoveries are there, whether we stumble across them or not. They continue to flower on different branches of the same tree.

> The social cognitive ontological constructs mouthful of detailing was necessary to appreciate how it maybe possible to create human emotions in artificially intelligent computer programs by simply figuring out a way to run two different computer programs containing two different social cognitive ontological constructs (written in code) in a manner that creates an acute friction or overlap or a locking of horns or entanglement of codes between the two.
>
> Sutra

Chapter 43

REALLY, A WHOOOOWHOOOOOOWHOOOOOOO AM I TIME, IT HAS BEEN

Looking at individual bodies it seems each one of us harbor millions, which behave as one. My oneness is part of an eternal million, which I once chanced upon in a garbage heap creating patterns, dreams for a younger generation. For what is youth but a deception of time? Lift off temporal curtains; let the universe be seen as one, young and old alike in a continuous run. Touch my hand and feel the millions reverberating, the perturbations generating. Did I create a stir, a non-local effect of my electron kissing your proton? Feel…it's the feeling of our millions, yours and mine…interacting vividly in all possible outcomes, in all branches and networks arborizing and electrifying, that one, which is in you and all others and me perched on this tree.

Sometimes we become obsessed with a particular branch of the tree and keep trudging through it in spite of all the brambles. We either bleed to death or use a knife to tear through the brambles or better still leave the branch and set off for other greener pastures. For millenniums have I wandered through this tree…Many a times have I been wiped out totally only to resurface in a clod of Earth, or an early morning dew drop melting in the rising Sun. Was it only for a solution, an optimization…that my journey began through this decision tree. If so, then there doesn't seem to be a complete one. It can't be complete like a jug filled to the brim. It has to flow and splash out into rivulets and streams, meandering through valleys and rushing past civilizations, which have a habit of cropping up near its banks. So dependent are we on solutions flowing ever so freely that we cease to remember the original problem in which they were conceived.

> Another fascinating thing about trees is that their entire life trajectory is visible in a single glance at their branching patterns (that represents every decision taken at various points of time by the program inside their seeds).
>
> I look forward to a day when humans will be able to emulate their transparency and anonymity at an individual level perhaps in the next generation of electronic health records.
>
> Sutra

Chapter 44

NETWORK TOPOLOGY OF THE NARRATIVE

The structure of this narrative is such that each and every page that represents a chapter is a major node in a scale free network, some chapters developing into dominant hubs from time to time in the overall topology of the narrative. The story line begins with Con taking the plunge with his notebook June into an Earthy material life. A plunge that makes him fall through life and is bound to end in death. Death smashes his notebook's hard disc into tiny bits and pieces. The nonlinear narrative as a result tries to pick up these broken fragments and brings out Con's journey through the human body in its tree like statistical self-similarity with the Earth and the universe, which exists in an atom. It portrays microcosmic interactions inside the human body at a macrocosmic level of day to day living on an Earthy scale.

Con becomes well versed with the anatomy and circuitry of the various intricate components of his machine but also realizes that it has developed in an evolving assembly line whose creators are long dead and nobody till date understands perfectly how the damn thing works. However there are theories, stories of atoms and molecules and their subatomic families regularly utilized to explain how semi conductor chips work inside our bodies. Throughout the ages these stories take on multicolored hues, theories on the nature of Con, Earth and self-evolving machines...the science is ever changing. In the midst of all this learning, which Con realizes is endless and ever changing, his notebook, the body of June, is afflicted by a deadly virus.

> One can't just get a seed to communicate with a tree. A seed needs to be placed on the ground and one has to wait till it grows to a certain height before they see eye to eye.
>
> Sutra

Chapter 45

FROM A DEATH RAVINE

"What have you named your baby?" An unusual question thrown at a professor used to handling impersonal, theoretical queries. Samsara glanced agonizingly at the student who did that and found a normal looking girl with blonde hair.

"Well...Samsara sounded embarrassed...we haven't been able to think of one as yet but there are a lot of options. My wife keeps making lists of them from the net and finally after I have gone through the whirlpool of names scattered round the world I watch it disappear down the drain rejected by my wife."

"Call her Juneli...its Nepali for the light of moon", the blond said and vanished like a shooting star in a dark starry night. The name stuck as Samsara watched June grow up every day in between his hectic schedule of lecture classes, Baul research and human management (which was less about human control but more about setting them free).

He watched her grow even in his dreams that were very peculiar unfathomable states. Every night he would put his body in the standby mode, remove his glasses and doze off to a theme of recurrent and yet varying patterns on his screen saver. On these occasions he would sometimes even reach a state of complete darkness, stage 4 sleep probably from all he could remember from his physiology days.

On such dark nights he would suddenly chance upon the moon shining down upon a flurry of activity in a moon market on the ground below. He looked at the ground, the red Earth of Sushunia, its rock outcrops dotted with patches of green, reflecting light of a surreal full moon. He could visualize professor Samsara fast asleep beside his wife Maya snuggled up to him. There was a lot of time before daybreak and another hectic schedule of lectures. Time here stood still in infinite space. Con woke up with a yawn.

"Hi, I am Con again speaking from the heart of darkness. I am presently engaged in an arduous task of climbing a tree. It wouldn't have been that bad if it wasn't as spread out as I am discovering it to be…So many different branches each involving a major decision. Right now there's this fascinating branch with *Apsaras* and *Nartakis* daintily dangling grapes with their lovely hands…The beautiful soft caressing hands of Maya, my wife who's snuggled up to me.

I climb further and find her in the arms of another man when I am at the lectures...Why that scoundrel of a newspaper boy who I used to admire so much!...Working 20 hours a day from 3:00 AM delivering newspapers to 500 houses, morning and evening in the neighborhood, also working from 9:00 AM to 4:00 PM at Samsara's own college as a physics lab assistant. I should never have been impressed by his attitude and invited him to tea...How could Maya do this to me? I must say this was a possibility I had never been paranoid enough to consider and yet it was always there nesting its eggs in this one branch. Maya looked very demure, child like, sleeping with her mouth slightly open. As he ran his hands over her face and neck and felt her carotids he remembered how they were instructed not to press both the carotids together while examining for carotid pulsation. It blocks the total blood supply to the brain."

This branch gradually led on to a dark sky with not a star or moon in sight. For as far as was visible he could only see darkness and Con realized he was back to square one...the death ravine.

Chapter 46

To Life Divine

Con chose not to climb up any further on that particular branch leading to darkness. He retraced his footsteps to where he had started. He tried another branch which again didn't lead anywhere except a confusing array of different branches that were definitely not statistically similar (fractal, that is-Sutra).

He found the more he hopped on to these higher branches, the thinner and sharper they became. Right now he could see a few branches in front of him that were blade like sharp and he was sure his limbs would bleed if he stepped on them but he suddenly noticed somebody standing at the tip of that branch... A beautiful crimson breasted barbet wearing thick red-rimmed glasses.

"Wow! I am sure that's June, I've got to meet her, talk to her, learn medicine like never before," thought Con. "Hey Joooooooon, Wait for me...I am coming up".

"Wow! This is sharp and I am bleeding already but look at her, with that smile on her face, she's spread out her wings, no she's not going to fly...she's just welcoming me with open arms. Wait...I am coming into your arms baby! Wait for me...I am nearly there."

A crimson breasted binocled barbet sitting on the topmost branch of a tree looking out into a blue sky sobbing gently.

"Con, June, why didn't you wait for me? I could have sworn you were welcoming me with open arms and only when I embraced you I found I was holding on to thin air. I still can't believe you're not there. I can't believe you disappeared as soon as I touched you. I couldn't even feel your body. I feel saddened but there's also this inexplicable feeling of bliss which mixes well with sadness...it's strange...it's...."

A different voice, high pitched, sweet and melodious. "Hi! Con, this is June here; no don't look around because you can't see me. I am within you. I became you the very moment we embraced each other. Surprisingly there was no explosion annihilating us, as I had feared there might be. I suppose it was your courage which saw you through, which didn't allow you to choose those other branches leading to the murky depths of hell (Schizophrenia…all clichéd, I am afraid-Sutra).

> As for the soul of medicine, it's easy to understand at least that is what my friend Mephistopheles thinks. "You master the microcosm and the macrocosm and in the end let things happen as it pleases God (? Con)."
>
> Sutra

Chapter 47

BUYING AND SELLING CONS IN A GLOBAL MARKET

Jhinuk kurote kato chall, Jhinuke ekhono neel jal
keo ki kakhano mone bhabe, jhinuk kuriye deen jaabe.

The professor was spouting Shakti, his favorite poet, all the while bending down on the sea beach marveling at the innumerable seashells reflecting the rays of an evening sun that was also shining on his bald vertex, which only his students could appreciate.

"Con has a peculiar hobby. He just freaks out on machines; one's with CPUs that move around, talk and laugh, appear domesticated or just growl or mutter-live in the wild. He also likes the ones which are stationary, like trees, gathering all the energy from the sun so that all other machines can function."

He paused to buy green coconuts from a passing vendor who happily started gouging holes in each of them followed by introduction of plastic straws and a lot of coconuts and money changed hands between the professor, his students and a visibly thrilled green coconut vendor whose day was made.

"Trouble is, Cons can't be bought as easily. You have to observe, get to know them better. Most of them won't even let you touch their lap top exteriors so that you could immediately get into the innards of their hard discs, read their "my documents". No way, for that you have to spend a lot of time, observing, listening, and questioning (most importantly asking the correct questions...hitting the correct thought keys).

Eventually, when you get to read their files its fascinating...To tell you the truth my life's calling is to simply go on reading their files, marveling at them but that's not half the joy of it really! It's the bonding, which you develop, with them, the trust they place on you, the memories you share on your files, cry silently together. For all that one has to possess a body, a notebook of one's own. However, the day Cons started possessing one; consumerism was born fuelled by desire and mass production. There was this added fear of death...the fear of losing their machines and all their personal memories on their hard discs.

If only they could change their machines without losing their memories or better still carry their hard discs with them all the while on re-writable CDs so that every time they could pop it in any machine within their reach and work within their personal digital environment. They have a peculiar fixation on their personal memories, which have actually evolved from interacting with memories of countless other hard discs connected to their network. Most of them would like to treasure their memories in their own machines or password protected personal web spaces. This attachment is what kills them, even before their machines have a chance to die. Why couldn't they just be...without having to possess the body of their machines? Couldn't they just control them with a remote going about their daily routine, all from a distance (chilling out somewhere in the Himalayas)?

They wouldn't feel much then...even if they have a violent sparking in their innards or some joker smashes them on the ground. At least they wouldn't fear dying as much as they do now. Dying would be another show ending on their monitors."

"Wouldn't it make them couch potatoes?" one of the students ventured. "It could also make them trees. I always felt trees were people who had reached that state, watching us on their remotes. For each one of us machines, there has to be a tree...our very own dear lil tree. There we go again back to personal possessions. It's difficult to have a permanent solution to this due to the fluidity of our existence. We are dependant on our machines for survival as much as they are dependant on us. We have a long way to go before we become trees or mountains or rivers. That too, if only we let them survive along with us."

They had arrived at a point in their coastal trek where the beach ended and a rocky cliff stood blocking the road with the sea dashing against it. The professor started roping up his students for another traverse.

Chapter 48

A STORY ABOUT THE BIRTH OF ONE

Dyspnoea is an abnormal awareness of ones own breathing. Normal awareness of ones own breathing is meditation. All of us must become abnormally aware of our breathing one day when we die. Being aware of it every moment everyday till then may just help to ease the process of death when it comes.

Sutra
A lecture on metabolic encephalopathy

A visual of three people being executed by being walled off into a brick house. After the last brick has been installed these people trapped inside experience pitch darkness and their LCD screens begin to zing bing with hypoxia...that's when our story begins. It's a story about the birth of a story. We begin our day with the romance of the story, falling in love with whatever work we begin and by midday we get married as work progresses at a feverish pitch and we feel hungry for lunch. Finally by night, death and sleep ends it all preceded by an evening walk and a visit to the library just before dinner.

Con wants to have a body and starts at a point of time claiming one such when it's in the process of being unclaimed. The LCD screen blinks and sputters. Strictly speaking, he doesn't need his entire body, only the part that would allow him to synthesize his perceptions into a film in Technicolor…the narrative visual imagery that requires so much disc space.

Somehow his entire body seems necessary to run the accessories, which as a whole support his preoccupations. Like having perceptions meant sensory inputs which if desired in a better quality would require him to move around a bit (he wouldn't give up on his evening walk for anything in this world).

So he accepted the fact that an input device with five sensory ports be fitted in as well as two pairs of limbs to cater to his motor movement urges. It would have been alright if the motor movements were confined to his limbs but no, there were bowels to consider that had a tendency to move at the most inappropriate moments or sometimes not move at all when he wanted them to, especially in the mornings.

He chose a human model not because he thought the exterior looked fetching or useful but all the thirty three crores of other species were already taken up by other Cons (according to dated Eastern mythology). It would be his very own notebook, a toy for a good number of seven decades (unless it had a manufacturing defect or was destroyed by a traumatic blow if not a bad virus or bacteria on a rampage).

Now what is it that makes silicon mother Earth reorganize into a notebook, a seed grow into a tree, an embryo into a hen or human or Orangutan for that matter? What is it that converts a string of words into a story?

What converts the string of words into a story is the sequence of the words... Right??
Jyoti Bhojwani, 17 May 2006:

Thanks, reminds me of the DNA governing our chemical individualities….Sutra

Chapter 49

WORDS AND IDEAS

> An exercise I have found very useful draws from improvisation drama: "The one-word story". If you have 10 or 20 individuals then ask them to sit in a circle facing each other. You then feed them an initial line of a story (e.g. "The girl went into the forest"). Then tap someone on the shoulder and ask them to add a word to the story. And so this continues round the circle with each person adding a word to the emerging story. I found this to be a good experiential exercise since during the de-brief participants say that no one individual was in control but a story structure clearly emerged. They also reach the point where the emerging story is influencing what is said next. Some talk of the futility of trying to control the story and how frustrated they were when the person next to them uttered a word they had not predicted and so had to respond differently.
>
> Boyd Rodger,
> www.learningecology.co.uk

It was a class trek again this time in the jungles of Buxa and Jainti in the Duars region of Bengal. There were 33 tigers counted in these forests in the last census under the tiger project. It was not however the tigers that intimidated them. It was the leeches in the rain. The professor started off, "Words are leeches quivering on leaf tops in a rain forest dying to attach on to any passing idea however unsuspecting. Ideas sucked dry to the bone make words fall off burping to their hearts content. Ideas are but poor innocent souls, roaming in the forest looking for a cozy shelter before night falls suddenly in a dense jungle.

Eventually they do find one, after losing the way umpteen times, inside a forsaken forest bungalow. They share words in the light of a candle, words like fishes, succulent and juicy leaving an after effect long after sleep engulfs our ideas.

Ideas sleep.... only to wake up at dawn, greet another sunrise of words and sentences amidst raging mountain torrents. Ideas are washed away into far off river valleys and flood plains where they slow a little. Children bathe in them with gaiety, sprinkling soft sands. Words are children looking for their long lost parents, homeless and yet full of vigor, determined to find their genetic donors, who exist in ivory towers, reach up to them for comfort, joy, satisfaction, security... what not. From time to time they run after these conventionally important things, which peculiarly hide in ideas, so vast and spread out that words are just a speck. A tiny grain of sand in comparison and yet these grains are like huge atoms, with their own solar systems and electron clouds. Space cities built into their infinite elementary particles are but one gigantic idea that sprang from an astonished contemplation, accidental and yet profound. It dwells in our minds, which long for a theory, a solution to everything, in its every day nothingness.

Words like genomic imprints, building blocks of complex molecules, keep changing rearranging themselves, generating more and more ideas, different cells performing different functions, diverse set of ideas, which like disease, manifest themselves on integrated complex molecules of life that we fancy ourselves, more often than not leading to death, obliteration of one complex molecule, any one amongst us perhaps, in the vast shores of time, another re arrangement of on-off signals. The network continues to function as if it knew no time."

That night they had roasted pork with beer lying, in a forsaken forest bungalow, singing songs and sharing poems under the light of a full moon.

Chapter 50

SMASHED TO SMITHEREENS...PICKING UP MEMORY FRAGMENTS

This is Sutra again...had to butt in as this part is a bit tricky.

All you can see in front of you is a deserted beach, which meets the Sunderban forests somewhere off beat in Bakkhali. As you start getting the creeps seeing all the Hental brambles and other trees proclaiming tiger country, you suddenly chance across a lot of silver shining sea shells strewn across the beach and a lady in her gown with her hair let down crouched over collecting sea shells.

You chat her up and get to know that these shells are nothing but what is left of the life of June. She was naturally smashed to smithereens after her fall and her personal narrative that is so much of an essentiality in this wider narrative of system trouble shooting, was now a smashed and fragmented hard drive like a shattered dream her life was.

This lady was an avid reader who was collecting these fragments for us (making our job easier). If it were not for these people the world would be so much more fragmented, in fact it would be nothing but stellar dust.

You might call them weak forces but in the grand unified scheme of things they do have a major role to play.

Chapter 51

LIKE FRUIT BATS IN THE MOONLIGHT GLEAMING

The modern day notebook is an individual with 5 sensory input ports and analog devices for outputs both somatic (like the limbs) and autonomic (like the bladder and bowel he needs to move everyday). Strangely this individual, this notebook, although it has a marvelous brain of its own...it can only function at the hands of a hypothetical character of uncertain origin, collectively labeled Consshness(Con for short).

Con in turn thrives on it. It's unimaginable how much Conly communication would have suffered if not for their notebooks. Just imagine a few Cons sitting in a train compartment, dreading the thought of starting a conversation for fear of having to look each other in the eyes. Its odd how they can't reply reflexly, they need more time to think on each of their outputs that they conjure on daily emails. The train compartment situation is still salvageable if the passengers hitherto silent flick open their notebooks and immerse themselves on their LCD monitors. If they are lucky they might find a way to track down the email addresses of their co-passengers, through a brilliant google that would oblige key words like, so and so train with so and so time and so and so seats, passengers' names and addresses.

After a time as the train picks speed co passengers would have started sharing files, well carved narratives of their past, (which would have been otherwise garbled at a painstaking attempted free speech). It would have been delightful to learn that the uninterestingly serious faced gentleman with a white beard was under the impression that their train compartment was the interior of a spaceship and his destination (where he fancied he'd reach next day) was somewhere near Rigel in Orion. He in turn would have been equally surprised sharing files with the young man sitting near the window with that faraway look in his eyes. He thought the umbrella like cemented pillars for the upcoming city flyover, their train was passing were giant mushrooms. "Funny, the old man thought, he could have sworn they were foundation stones for a new hyperspace bypass." The young man was already looking forward to the time he'd shelter under them in the rain and with each pelt down the mushroom would bloom bigger and bigger until it covered the whole sky and sheltered a whole new generation of aggressively pacifist poets with their side bags containing battered diaries that they would pull out from time to time spouting poetry like humming kettles (unlike having to share them on emails like the present).

However barring a few exceptions like this young poet, most Cons wouldn't cater to sharing thoughts on speech through an air media via laryngeal sound generation and cochlear reception, the cochlear fimbriae rustling and swaying gently like paddy in a breeze. They generally would prefer the email mode of communication. The chief reason being the volumes of data that could be transmitted on emails if tried on speech would make the larynx give in after a time. That is why the ape-man (again under Con control) graduated slowly from the primitive 486 into the modern day Pentium/AMD and learnt to read and write. Before that Cons could only think of the abacus, none of the modern day human wonders they clutch at most of the time hanging onto their shoulders like fruit bats atop acacia trees gleaming in the moonlight.

However to come straight to the point (which nearly always has the danger of being lost in between the lines), I would like to point out the plight of our index Con who wants to know more about his machine but to his dismay finds that no one really knows the whole story as it's lost (very much like the point in a line and a line in a plane, a plane in fractals and so on) in a multitude of broken narratives.

It was as if someone had carelessly tossed away the fragile story line making it fly for some distance before it hit the ground and was smashed to smithereens. The story of the modern day notebook with its arms and legs and mouths and anal orifices (correction-there's only one mouth and one anal orifice-Sutra), is that of a unit like a cell forming the body of the Earth with its intricate branching and circuitries of trees and rivers that are visualized once the lid of the sky is lifted off (Aren't we repeating this too often?-Sutra).

Our index Con needed to know the story of how the universe was created with Earth like a cell among countless other landmasses in its bludgeoning galaxies. Imagine! Just because he needed to know how his notebook worked in the first place!

Chapter 52

A Fragmented Con in Small Things

Con was trying to see beyond what had been allotted to the machine, which he had bequeathed. Unable to see anything there, he looked back fondly to the days when he could see all, that is, before the lights went out.

When they reappeared he was staring through the optic disc of a moving notebook with hands and feet that could feel its nose and ears, which in turn could smell and hear. Interesting input devices but wholly inadequate for taking in the enormity of what he wanted to behold. The trouble is Cons themselves had fragmented over the years and taken ownership of specialized systems to cater to their own needs of the moment. The result was a lot of Con in small things giving a multidimensional perspective so much so that the need for dimensions went on increasing.

Gone were the days when there was a single limitless dimension stretching into the infinite. Everything in the microcosm and macrocosm would be visible, tangible at once, like you could see all the electrons and protons in an individual whom you were chatting with and could very well say Hi to the person sitting behind her. Also you could see through walls...those were the days, except that there wasn't anybody in particular to see all that because Con was un-fragmented and living machines were yet to evolve. They were yet to develop their eyes, ears and even limbs that could touch and feel. Things were simple and Con was happy but then when material things began to evolve from simple cells joining hands (oops! cytoplasmic projections) matters became uncomfortable so much so that in the present age Con has been left in the dark.

He can't see without using a machine. He has a wide range to chose from, plants, animals, humans, not to leave out stones and minerals (so called inanimate machines that really have a lot of animation at the microcosmic subatomic level). Off course, none of them could see all like Con could, when there was no one to see in particular. So our index Con gets stuck with a human notebook, the walking and talking kind and generally has a whale of a time sending and receiving emails, his window to the world. He had great fun waiting for those emails. They contained loads of useful stuff that one could browse and ruminate and reply to after a lot of contemplation. He would copy paste them again to a number of friends taking care to make them feel, as if it's been addressed solely to them.

Con thought all those chatting soft wares were rather primitive where one had to reply instantly and reflexly without much contemplation. It was all right for Cons who were running the rat race but not our index Con about whom the story is all about. Each and everyday they kept him busy because he was supposedly the most knowledgeable fellow around who could fix things, not just plumb their xylem and phloem blockages but also coax photons from the sun to release electrons from their cytochromes, particularly when the sky was overcast. He fancied himself as a mediator in the overall scheme of things, the spontaneous creation and evolution of which was a mystery to him that seemed to deepen with each passing day as he saw, knew, realized more, he sensed the enormity of what was still un-chartered.

Chapter 53

IDENTIFYING AND PLUCKING LEAVES IN THE EVIDENCE JUNGLE

He had been living with these complex machines for long now, from the very day he had answered the medicine professor's request in the affirmative and had acquired a brand new notebook all for himself, straight out of the assembly line, a huge factory with cells forming regularly one out of the other, subsequently being sent to separate compartments to perform specialized functions. He had tried observing the process of creation in the hope of learning the whole of it. Yet the process remained mysterious as the Earth where all machines were trees, rivers, mountains, animals and humans interacting, evolving and changing the body of the Earth so rapidly, that Con felt its shelf life couldn't be a few more million years...sometimes he wondered if trees, rivers and mountains could be mechanical systems at all...only when he peeped deeper into them at a microcosmic level was he convinced of the enormity of subsystems and the complex interactions each of these systems held. None of them were simple but looked beautiful from a distance, like the Annapurnas.

Durga Devi was a middle aged lady, who'd come into his OPD, a place where machines with mild to moderate dysfunction were wheeled in for repair. Con was in his mechanic's overall and as both doctor and patient interfaced each other, Durga Devi saw Juneli (the machine which Con had been using all these years of his mortal existence to interface with), a girl with plaited braids and a pleasant smile through those thick glasses. Durga used to play the harmonium while her husband sang, both having been trained in Indian classical, a learning, which had been used all these years to enthrall a small assortment of friends, neighbors and well wishers whenever they'd manage time from their not so busy nine to five jobs in the secretariat as clerks.

Lately she was finding it difficult to follow her husband with her fingers while his voice moved nimbly on its vocal chords (even at a ripe age of 69). Irritation had given way to alarm and she'd sought June's help. That was 3 years back. Her right hand since then had become stiffer along with her right leg and there was this inability to hold her trunk, stand straight. Things weren't this bad till 2 months back. June remembered how Con had plucked leaves of Levo dopa from the jungle (*Levo meaning one which rotates to the left when the wind blows from the right--that's how one identifies leaves in a jungle-Sutra*). They had been lost in the jungle once picking leaves and were nearly swept off their feet trying to cross a mountain torrent following which they had gotten drunk on country liquor while the sun set in that 33 tiger infested jungle.

All that effort came to naught after all these years. Her stiffness seemed to be worsening with each passing day even on levo dopa.

The more June learnt about diseases the more puzzling they seemed. Why did the Levo dopa in Durga Devi's brain dry up in the first place? Why did the effect of Levo dopa (collected so very painstakingly from the jungle) and administered externally into the blood stream wear out after a few years?

Chapter 54

PRINCIPLES OF EXTERNAL-INTERNAL MEDICINE

There was a particularly interesting shell sparkling on the beach. As June stared at it she could visualize the bearded face of the professor standing on the pulpit addressing the students. His narratives were always so very..."Today I am going to take you all to my retired professor, Dr Joatmon who lives in the outskirts of the city. He's from a time when troubleshooting humans was not as compartmentalized into specialties as it is today. Even now, he tackles a few of our patients when we are stumped."

The car stopped on the road that had crossed a dense jungle. June couldn't believe it was just a two hour drive from their college. She treaded sheepishly on the foliage half expecting to be bitten by something but was partly reassured by the professor's confidant strides.

They crossed a few rivers and the last one left her drenched from head to toe. The current nearly swept one of the students off and another lost one of his shoes. They had to climb down to their last river, pretty technical at places. Thank Con! For those fixed ropes...wires actually, some pilgrims may have left.

Then there were those slippery steps cut out from the rocky ridge and as the professor clambered up them, June sat on them to dry herself. After sometime she found a gentleman with a white flowing beard grinning at her. She'd have taken him to be another version of the professor if she didn't find the professor accompanying him with an expression suggesting a 6year old school boy who's just found his pencil box hidden someplace completely forgotten earlier.

All the students stared awestruck at Prof Joatmon, founder editor of Joatmon's "Principles of External-Internal medicine"." Hi! I am Joatmon call me Jo for short. Come on up for a cup of tea." It turned out to be a thousand more steps, not that June cared to count. They were sipping tea in a makeshift home under a huge stone slab labeled, 'Jo's Cave'. In front of them was another dark hole. “That's the Mahakaal cave, "Jo said in a matter of fact manner as if June had asked the question in an equal vein. It was discovered by one of those old traveler type missionaries. Its all limestone inside with a lot of stalactites and stalagmites, which the local people here venerate as Shiva...you know, the penile figurehead". June wasn't sure she'd need all this info but professor Joatmon was a strange man... a jack of all trades, master of none was what he said his name meant unabbreviated (if you took it to be an acronym).

Medicine is all about being just that he said; after all it is Con only who can master the human body (and there's no point mastering just a part).

Con pricked up his ears at the last statement. "Well here's one fellow who's finally talking sense.

Chapter 55

ADD-ONS TO CONSCIOUSNESS

Prof Joatmon began, "To understand system malfunction it is necessary that you understand system function for which you have those pre-clinical years where the intricacies of system components and their functioning are detailed. However there's more to it than just knowing what's inside the body.

Equally important is what's outside and how the machine interacts with its surroundings. Like, why do we need our machines at all...it's to have fun, it's to revel in those five senses, process them creatively and express oneself in motor activities that trigger passions generated in neighboring machines. In short, we have fun unwittingly, only to run a bigger machine, which is the universe that also does the same with other neighboring universes playing inside that infinite one.

In medicine one tries to look at the infinite like looking at the whole forest and not just concentrating on its trees." Joatmon's eyes had that dreamy look. "Trees...trees everywhere you look into the anatomy of the Earth. You actually need a microscope (or telescope...call it whatever you want), to peer deeper into its tissues...the so-called microcosm. Trees are receptors getting up or down regulated from time to time, homing variety of molecules, birds returning to their nests at dusk. This is just a small peep on lifting the lid off the notebook and you sure will need to know more as time goes by and you keep peering into these systems in the hope of trouble shooting them.

After you've lifted the lid off the Earth and studied the microcosm you'll need to study its behavior in the macrocosm. Like for example what is its idea of having fun...the very purpose of its existence. There are Con's with machines (people) who simply exist for a daily struggle against death and deprivation, their fun lies in getting to fill their stomachs each day after obtaining the means to fill it. There are Cons who continue to use their machines (bodies) in spite of their limited functionality (due to disease or overuse trauma).

These are our day today patients, machines whom we try to repair, restore to normal function or simply add supportive devices so that they just continue to live. A 20-year-old girl had come to meet me today. She works as a receptionist in a 5 star hotel and both her kidneys have been destroyed in a forest fire. She's learnt to use her peritoneum (the bag all of us carry inside our abdomen covering our viscera-Sutra) to clean her blood, not as good as her kidneys but at least she will live with it. Every day she needs to introduce 4 bags of hyperosmolar fluid into her peritoneal bag and hopefully most of the harmful molecules normally filtered by the kidney would filter into her peritoneal bag. (*We still don't know what all these molecules exactly are but we have devised a system to take care of them...can you beat that! –Sutra*). In this manner she'd throw away the wastes in her blood along with the hyper-osmolar fluid that she'd need to hold in her peritoneal bag for a few hours everyday. Hopefully after a few months of training her daily manual filtering would become second nature, like brushing teeth.

Most Cons would do anything to just continue living; continue to function if even with artificial support. All of them are very attached to their machines. They hardly ever realize the machines are simply add-ons to their consciousness (themselves).

Sutra

Chapter 56

A MEDIAEVAL SPECIALTY CLINIC

"Jo how do you manage to run so many specialty clinics single handedly?"
"It's possible if your specialty is life."

"How do you investigate them?"
"It's simple really! I just start from the basics.

For instance nephrologists would garner a lot of information by taking a good look at the urinary sediment. All that debris the riverbed gathers on reaching the plains. Right from its trickling down from the glacial filters of the glomeruli to the tubules, collecting ducts, ureter, bladder, urethra and the sample mug, it's a journey of filtration, re-absorption and secretion of the body's metabolites, molecules in motion, like hordes of people rushing out of the North Sealdah station to catch the south bound trains.

Microcosmic interpretations at an ideal plane would consider knowing all about these people but then it's difficult for a single person unless you have a good information network storing data from the hard discs of each of them.

For the time being we have to make do with what we see from a distance, from the tip of the flyover opposite Sealdah; Hordes of people, walking, jostling, chatting, quarrelling...quite normal you would say but in times of renal injury to the glomeruli or tubules you find the market places deserted, a few shops or trees burning, scampering feet followed by police batons.

Chapter 57

THE MEDICAL GROCERY SHOP

June is in her teens now (*as far as this episode is concerned-Suts*). She's a 3rd year medical student and a regular member of Professor Jo's class on patient centered medicine, which she's quite enthusiastic about along with her regular medical college routine. She listened as Joatmon began with that faraway look in his eyes.

"As physician healers our only job is to trouble shoot advanced systems and there is a physician healer within all humans irrespective of their technical qualifications. To utilize one's healing capacities one has to learn to be a life long student. The stories we experience everyday tell us how to approach various system problems and how they are solved. Essentially they portray disease situations where macrocosmic expressions of these systems go awry due to peculiar microcosmic interactions."

Having said that professor Joatmon attended to his next customer. He was at the moment a shopkeeper with an aura of a physician. He was a shopkeeper who liked chatting with his customers, listening to their problems taking care not to offer immediate solutions. June thought he resembled the huge tree she had spotted once in the midst of the marshes in Bharatpur his patients perching on him each day like a host of painted storks building nests on his shoulders. He never made an attempt to shake them off but simply filed them in his data-base.

Joatmon began again after having disposed his second customer for the day. "These microcosmic interactions would be different in different individuals as all of us carry within us a chemical cauldron hidden deep in our bodies, the chemicals brewing within our blood. Different molecules with different expressions give rise to our chemical individualities.

In our approach we don't treat each and every machine as if they are created equally. We don't type them into disease classes, rather we look at them as individual systems afflicted in a manner that is unique to that particular system. Each and every machine has its own way of adapting to its immediate environment although superficially they may share a statistical similarity. Inside each and every machine if you look into their circuitry is a tree with a pattern of statistical self-similarity (that is very often labeled as a fractal).

A fractal structure is present in all our systems, be it a river like branching of capillaries and blood vessels seen on Endoscopy or a bronchial tree on the chest x-ray carrying air to the alveolar rivers for supplying oxygen to the cells. When disease strikes due to a particular machine's maladaptive behavior, the fractal tree is chopped off or scorched, scarred and distorted as a manifestation of the battle within."

The next customer had come for half a kilo of sugar. The grocer professor gently filled up the weighing scales with crystals of carbon glistening as sugar, all the while enquiring about his family, his wife's headache, if he could put his little one to school this time or had the principal failed him again in the parent's interview.

Chapter 58

QUEST FOR CHEMICAL INDIVIDUALITY

> "To our chemical individualities are due our chemical merits and shortcomings. The factors which confer upon us our predispositions to and immunities from various mishaps which are spoken of as disease are inherent in our very chemical structure and even in the molecular groupings which confer upon us our individualities and which went to the making of the chromosomes from which we sprang."
>
> AE Garrod

June watched and marveled at the professor's unique flexibility, his ability to work at many different levels simultaneously and establishing individualized care rather than a blanket protocol. Protocols and decision trees were what June spent most of her days memorizing in her medical school and professor Joatmon's approach struck her as peculiar. At the same time she found this easier and less taxing and began to seriously doubt her medical school texts that treated all patients with one diagnosis in the same manner. "Don't you find this ironic?" She asked her batch mate and best friend Krish (shortened from Krishna). Krish was a tall dark handsome guy frequently spotted with June in the canteen.

"What?" Asked Chris, somewhere down the line Krish had been nominally transformed into Chris). He was trying to concentrate on the material for the community medicine exams, trying to memorize the dimensions of a sanitary toilet along with mugging up a lot of info (facts and figures) on strategies for disease prevention on a global scale. June continued, "Well, don't you think it's ironic that a profession so very devoted to the care of individuals should care so little about individuality."

As a third year medical student, June wouldn't bother to wait for the bedside teaching to start everyday. That was unlike her other batch-mates loafing in the corridors, presumably waiting for their professor to allot them a case, who would very often be found at the coffee club, busy discussing the latest bit of evidence on the efficacy of a newly launched drug. June would rather allot herself her own cases for the day by simply walking up to any of the individuals lying in their beds waiting for deliverance. "Hi! I am a student doctor and I want to know what your problems are?" she started off with a particularly emaciated man with a distended belly.

"I am hungry." he replied.

Chapter 59

LISTENING TO PATIENTS AND TEACHERS ALIKE

June realized that a systematic step-by-step questioning approach wouldn't lead her anywhere with this patient. She decided to jump tracks like she had watched Joatmon do on a number of occasions.

"Where are you from?"

"Nepal."

"Well...I am Juneli..."she began and was cut short by the patient who started excitedly in Nepali, "*Aae...tapaiko ghar kahan pareo*?"

June explained she wasn't Nepali but her father was a Nepal lover who had chanced across this river in Nepal that gave her the name. Ram Hari opened up better after that.

June checked if she had got all the components of the history of present illness in order. The onset and duration of symptoms, course and present status of the illness, the investigations done and treatment received and finally noted specific symptoms that would help to localize the disease to any organ system component.

Samsara had once told her about one of his wild schemes wanting to convert these points in the history and physical findings into weighted numbers. "Let's assume each of the points in the history are numbers like a1, a2...a'n', b1, b2...b'n' for the general survey findings and c1, c2, c'n' for the systemic examination. If we integrate them for each individual patient we can have a number for a diagnosis."

June knew what Joatmon would say to that, "By giving everyone a number we assume that all individuals with numbered symptoms are similar in their inner microcosmic interactions but they are not."

Professor Mechanic was scheduled to take their clinical bedside for the day. He preferred that classes even on practical clinical approach to patients be held in the classroom next to the outpatient department, as he didn't like too many students standing and crowding around a patients bed. Also if the students were seated and relaxed the discussion was more enjoyable.

Only, it had to be labeled as bedside teaching anyway. The student who had taken the history began...so and so from...came with the chief complaints of abdominal pain and gradual distension since the last few months. The professor along with the whole class listened patiently to the complete history narrated by the student presenting the case with minimum possible interruption. At the end the professor asked one of the students in the audience to analyze it upon which he looked bewildered. As if anticipating this, the professor proceeded to explain it himself.

Chapter 60

A Young Alcoholic Cirrhotic

June would visit Ram Hari every morning and collect his story bit by bit during the time most of her friends would busy chatting away waiting for Professor Mechanic to come and teach them. Ram Hari was a 25-year-old man working as a hotel waiter in Bharatpur, Nepal and had come to Kolkata, India searching for a job in a bigger hotel. He landed up in the bed before June however as he had developed a distension of his abdomen while in Kolkata and when he presented to the outpatient department of their medical college a kindly physician admitted him. That was three years back and ever since he had been depending on the hospital for his food, shelter, and intermittent fluid removal from his abdomen. Most of the hospital staff knew him by name.

He had been a regular drinker since the age of twelve. His mother had abandoned him when he was a 6 year old (his father kept saying she had run away with another man). From the age of 12, he was forced to work as a daily wage earner by his father and was employed in a factory manufacturing country liquor. Since then he started drinking 250 ml of spirits daily and would occasionally consume 2 bottles of 750 ml each. At the age of 21, he suffered an episode of massive blood vomiting and tarry stools for which he had been admitted to a district hospital in Nepal where he received 6 units of blood transfusion. Soon after discharge from that hospital he resumed drinking for another year until he was readmitted to the same hospital with a fluid filled belly. He has undergone repeated removal of fluid ever since with needles jabbed at various sites over his belly He also had a hernia over his umbilicus due to all that fluid in his belly and which burst spontaneously one fine day for which he again had to be admitted for an emergency operation.

June started doing a physical examination on Ram Hari meticulously elucidating the findings as she had often seen Joatmon do.On general examination Ram Hari was visibly very pale and malnourished, had a prominent scalp infection, and his eyes suggested a mild jaundice. Ram Hari's abdomen provided the maximum number of findings. It was distended with prominent veins and fullness in his flanks, an everted and bulging out umbilicus (as if someone had stuck a top over that portion of his belly). She could hear a continuous to and fro sound over his umbilicus with her stethoscope but couldn’t figure why there was this sound and decided to look it up later in her books if possible. June could palpate his liver and spleen by dipping her fingers into the organs through his fluid filled belly. She could elicit a vibration of fluid over her palms placed on his belly by tapping at it from a point opposite. It is a sign that goes by the name of 'fluid thrill' possibly given by a thrilled physician first to have discovered it. It confirmed the presence of a lot of fluid in Ram Hari's peritoneal bag covering his intestines.

June took a peek at his file to look up the investigations already done during his hospital stay.He had a hemoglobin of 5 g/dL, a total leukocyte count of 1800, and markedly reduced platelets on smear suggestive of hypersplenism. His serum bilirubin was 3.6 mg/dL with an unconjugated fraction of 2.4 mg/dL, serum albumin was 2.1 g/dL and globulins were raised at 4.5 g/dL. His HbsAg (Virutex/latex), Anti-HCV (dot blot Assay) and HIV (ELISA) were negative. Ultrasound abdomen showed an irregular liver surface with portal vein measuring 14 mm apart from free fluid.

Chapter 61

BEDSIDE TROUBLESHOOTING: A SYSTEMS APPROACH

Prof Mechanic began, " The approach to clinical problems (or challenges), begin almost always in trying to localize it at a component level, be it in the kidneys, heart, brain or liver. Asking where the problem is gives us the morphologic/anatomic diagnosis. The ability to make a diagnosis would therefore depend very much on our knowledge of anatomy, the intricacies of the machine's internal circuitries. None of you should feel uncomfortable looking at the jumble of chips and maze of signaling pathways that one finds on removing the lid off your machines. You have had sufficient time I hope to learn about it in your pre-clinical years. This learning and your knowledge of anatomy is what separate you from a layperson that is more likely to be your patient. It is also important because the human machine takes a long time in its evolutionary sequence to change in structure.

So for most part of your medical career you will find many of the other things that you learn today shall change dramatically over the years but your knowledge of structure/anatomy shall not. Even grossly your machine may not appear to change very much in function at a component level. The kidney will continue to produce urine, the liver bile, and the heart shall continue to supply blood and oxygen to all the tissues, microchips that produce energy from your daily bread and the brain shall continue to think perhaps! This view of the brain was not always there in the past and we have held on to it for quite some time now. However at a deeper microcosmic level the stories may change. This is because, (Quoting Blake) "*The Microscope knows not of this nor the telescope: they alter the ratio of the spectators Organs but leave objects untouched.*"

Apart from identifying which component is malfunctioning we have to also answer the question of why it is malfunctioning. If a machine points out or displays chest pain on its LCD screen we can speculate the source to lie in the coronary pipes which may be blocked and not supplying enough oxygen to the central pump, or it could even be something wrong with the aortic valves in the pump, again impeding circulation through the coronary pipes. We can also speculate that it may be due to a host of other system components like the food pipes or lungs or even the muscles or bones of the chest and very often these possibilities would be narrowed by weighing clues offered in the symptoms, the onset, duration, course and exact site of the pain. Assuming that we have narrowed down the problem site to the pipes supplying the heart (fashionably called the coronary arteries) we have to wonder next why, what is happening there?

The commonest story in circulation at present to explain blockage of these pipes is the story of atherosclerosis. The story of immigrant lipid moieties settling down in the endothelium, constantly under attack from inflammatory mediators until one fine day the plaque like community fissures and cracks up resulting in a blocked pipe. This is a problem of the present global community, violence and strife everywhere. Our diagnosis generally ends here after unearthing the site and cause of the problem.

Chapter 62

SUMMARIZING LIVERS AND RIVERS WITHIN

Prof Mechanic continued, "To summarize then, we have a 25-year-old system with a bloated belly, past blood vomits and altered behavior. This point's to the liver and its pipes under high pressure as the anatomic location of this problem. This is what we would label as its anatomic diagnosis. Now the exact cause for injury to the liver is also quite clear. Ten years of uninterrupted high dose alcohol consumption, like spilling coffee everyday on your notebook that finds its way into the hard disc.

On examining this machine we find prominent veins over his belly a marker of all those displaced liver rivers. And just look at those beautiful red spiders, Wow! Look at that haul on his back! Look how they blanch off and disappear when I press this glass on his skin and reappear when I remove the pressure.

Also if you remove his underwear and examine his genitals you shall find both his testes small like that of a kid. There is hardly any secondary sexual characteristics in this 25-year-old you would expect, axillary, pubic hair...hardly any? The spiders-dilated small arterioles, small testes, lack of sexuality is all because of an excess of the female hormone, estrogen in his blood…That estrogen which couldn't be utilized by his ailing liver.

The class looked at Professor Mechanic mesmerized by his deft handling of this 25-year-old young machine arriving so very easily at a diagnosis both morphologic as well as etiologic.

"But then June thought, is getting the diagnosis the central process of clinical medicine?"

The fifth edition of the oxford handbook of clinical medicine that she carried about in her apron most often to the wards also raised the same question and in its own answer submitted, "The central process in medicine are: The relief of symptoms, the providing of reassurance or other prognostic information and the lending of a sympathetic ear, all seriously lacking June thought in professor Mechanic who always seemed to be in a hurry doing procedures, giving lectures and advice too often without much listening. He was more like a tree swaying gregariously in a storm shaking off all the birds that dared not build nests on his shoulders.

Prof Jo never seemed to be in a hurry, which made you wish you could be with him always.

Chapter 63

DISCUSSING THE WHOLE

June emailed Ram Hari's story to Joatmon particularly because she was convinced there must be another level of diagnosis for him apart from the fact that he had cirrhosis of liver (Prof mechanic's morphologic diagnosis) due to alcohol (His etiologic diagnosis).

Dear June, Thanks a lot for this. There is this other dimension of a patient's diagnosis which you have sensed and we in patient centered medicine feel is so often neglected by professor mechanic's hardware school. It's the social diagnosis that lifts this machine from the hospital bed and places him in a wider community perspective. That is why a hospital physician who never practices in the community can hardly ever connect to his patients beyond the mechanical approach, barring exceptions. I would like you to prepare a report of this case because it's important that we tell the world that cirrhosis with all its complications due to alcohol can even occur at this age possibly related to poverty, child labor and easy access to alcohol and we need to keep our eyes open for this cohort of young alcoholics. Give me the complete clinical summary of this case and I shall prepare the discussion. Jo

June browsed through the discussion prepared by Jo. It must have been hard to collect the references sitting in his cave in the jungle but then he had friends in better places that would have mailed the references to him. This patient highlights the social problems posing substantial challenges to health care in the third World. Child labor was forced upon him at an early age and he took up alcohol early also because of easy access to liquor. Inspite of legislative enactments prohibiting employment of children below the age of 14 years (section 14; Civil rights act, 2012, Nepal), the 1981 census, Nepal, showed that 4.5 million or 60% of the child population in 10 to 14 age group of Nepal is economically active1.

The sale or offering of alcoholic drinks to persons below 18 years of age is illegal according to legal restrictions on production, sale, and consumption of alcoholic beverages in Nepal. However, alcohol use has been quite prevalent in Nepal since time immemorial. Social tolerance to alcohol use is quite high and mostly alcohol has not been taken seriously either by the Government or by any social organization. Production, sale, and consumption of alcohol are ever on the increase and it could be taken as the number one problem drug in the country. These factors contributed to the early onset of heavy drinking in our index case and he developed cirrhosis at the age of 21 years. His dependence on alcohol was such that he didn't stop drinking even after the first episode of blood vomiting and tarry stools. Continued drinking for another year made his social and financial supports give away.

Following this he was converted into a total dependant on the hospital partly for his regular abdominal fluid removal and partly for 2 square meals a day, effectively falling into the vicious cycle of poverty and social-physical debility. The so called Developed countries such as United States have dealt with their alcohol problem with periodic enforcement of strict prohibition like the one, which lasted from 1920-23, and the incidence of alcoholic cirrhosis reached an all time low. The death rate from the condition fell to half its 1907 peak and did not start to increase again until the amendment was repealed. The time is ripe for Nepal to embrace strict enforcement of its prohibition laws.

Chapter 64

The Pump that Runs our Rivers

The 5th semester clinical medicine posting was the most exciting time for June with the days going by like a dream. Chris was being extra nice to her as a compensation for being not so nice to her father Samsara whose Baulish ways were getting on his nerves.

Professor mechanic's classes went on full swing. "System trouble shooting approaches a patient system by breaking it up into its component parts. Now that you have had some idea of how we deal with the liver we shall move on to the heart. It's vital to have a pump delivering a constant supply of electrons to run our microchip tissues. This pump that runs our rivers gives the required thrust to the electrons contained within them to reach the interstitial ocean. The ocean of water and its electrons exist not only amidst all of us cells but also within us. All of us individual Earths are made of three parts of water and one part land. In this three-part water of Earth's body there exist plenty of micro-capillaries, arterial rivers bringing our microchips daily bread and at the same time venous tributaries drain their daily excrement. A drain that returns all the water back to the central pump that purifies it by sending it into the lungs, open air, pristine valleys.

The machine right now before you is malfunctioning because it is unable to breathe properly. Its activities of daily living have been severely restricted as it gets short of breath even while going to the bathroom. Initially it wasn't this bad when the problem started. It was just that its Con noticed the system running slow, files taking longer to open as if fatigued. Then the shortness of breath was noticed on climbing stairs and now this. On general survey you can make out the increased breathing rate and also notice the veins near its collar becoming prominent with a peculiar bulging followed by a collapse, the V-y we call it. Looking at this we can safely assume one of the right heart valves is leaking and this in turn is because of increased pressures in the Lung Rivers leading to a pressure rise in the right heart chamber that has resulted in the right heart valves to come apart. Why has the pressure risen in these rivers, which normally are placid at 5-10 mm of mercury? Could it be an avalanche somewhere up in the snows, a lot of ice melting all at once? From the history we know it was a more chronic process.

The pump in his left side that presently drives the whole system circuit has been failing gradually for some time now. As the left chamber of the pump fails there is an increase in its end diastolic pressure that is in turn transmitted to the pulmonary rivers, which drain the right chambers. As a result the right pump fails because of the failure of the left main pump. So it's not just the breathing difficulty. His whole body is bloating up as the ocean in his interstitium stagnates for want of proper drainage. Now why did the left main pump fail in the first place?

It could have been a blockage of its own blood supply due to a sudden traffic jam in his coronaries (quite a common reason actually). Or it could be that the valves in the left main pump are leaking in which case you can easily hear that whoosh-whoosh torrential turbulence with your steth or it could even be that the muscle cells running the left pump with their contractile energy just dried up due to a poorly understood reason."

> Through it, over young women's tense abdomens,
> I have heard the sound of creation and, in a dead man's chest, the silence
> Before creation began.
>
> The Stethoscope by Dannie Abse,
> http://litmed.med.nyu.edu/poems/stethoscope.da.html

Chapter 65

BEHIND A FAILING PUMP: THE AILING PERSON

June followed up professor mechanics patient with the failing central pump after he got admitted into the ward. She worked on collecting his life event data that she was finding increasingly fascinating thanks to Professor Jo's approach along with the equally fascinating mechanisms of disease thanks to professor Mech.

The man with the failing central-pump with slowed rivers and a flooded interstitium was originally from a village near Sirmour in Himachal, India. Dharampal, that was his name, had three mothers from his single father. His first mother wasn't his own but gave him two elder brothers. His second mother was his own who bore only one child from his father and that was he. This was likely due to the fact that she died of a febrile illness when he was 3-month old.

He spent most of his childhood with his first mother after his father married a third woman who gave him 4 more brothers. He never went to primary school but spent his time cutting grass and grazing cattle in the valleys that were in the lap of the Himalayas.

During 1952 when he was 23 years old he went to get enlisted in the Indian army. He participated in the Taken hill fight near Mismi, Arunachal, something that June was unaware of having ever taken place. It was not mentioned in her school history text books. He said their strategy for taking over the Mismi tribals was to gun down a few, inflicting non fatal wounds so that they could later offer them first aid and win them over.

He was a great athlete and footballer from what June could gather from his account and when his unit moved to the Nathu La pass in 1962 against the Chinese aggression he missed all the action as he was playing league football for his army team as a captain. In 1971 he spent his time in the Indo- Bangladesh war against Pakistan in airdropping supplies for the Indian forces. In 1977 he was sent to Sri Lanka as a part of the peace keeping force and had road duty in Trincomalee, patrolling and clearing roads. In 1980 he returned victorious to his village. He had been never hospitalized in his 37 years of service and this was his first hospital visit. He had been married at the age of 27 years, the girl having been arranged by his first mother. He used to visit her in the village once in 2 years and she bore him 4 kids.

However in 1976 he married again, he said because he was irritated with his in laws. His second wife was 32 year old whose husband had deserted her. He continued to live with his first wife and clarified noting the look of alarm in June's eyes that both his wives were quite ok about it and in fact even after a year of his second marriage his first wife bore him another son. His second wife bore him 2 more children in the subsequent years.

Dharampal had already recovered from his shortness of breath before June met him in the wards and collected tit bits of information from his life event data in between examining him daily. Professor mechanic had confirmed his hunch of the morphologic diagnosis being in his left pump chamber by doing an ultrasound-echo examination in the out patient department itself. He also concluded that the problem was a blocked pipe to the pump, as particular walls of his left pump chamber were not contracting properly. Dharampal went home with only a mild residual breathing difficulty. The pump chamber pressures had been relieved with drugs that removed fluid through his urine. However he came back a week later with severe chest pain and breathing difficulty and died shortly in the ICU. It had been a sudden complete blockade of his pump pipes in all likelihood due to a sudden rupture of the cholesterol dirt plaque inside his pipes. Professor Mechanic concluded.

Chapter 66

SYSTEM TROUBLE: A YELLOWED DISPLAY VIRUS

June was regular in attending Professor Mechanic's clinics as she found his mechanical approach strangely exciting. Inspite of her beliefs about holistic medicine, strangely she didn't find the two incompatible. Even Chris pointed out, "Look machines are not really similar and have an individual adaptability so even if we approach humans as machines we need not fall into the typological trap of equating all humans as one. Human bodies can still be thought of as machines that need to be approached individually. Most of our problems stem from the fact that we can't separate human consciousness/Con from their machine like bodies."

At professor Mechanic's liver clinic June not only saw more machines with fibrosed livers with blocked inflow pipes she had found in 25-year-old Ram Hari but also a lot of other interesting problems like pipes bringing bile from the liver into the intestines getting blocked and bile spreading into the blood circuits as a result coloring the display a golden yellow. It was fun watching professor Plumber introducing wires with balloons into those pipes and removing stones and muck or putting in smaller stents (pipes) into the bile pipes if the block was due to those obnoxious looking schizoid growths many machines seemed to be increasingly getting afflicted with as age advances.

There were other causes of the LCD going yellow due to bile in the blood circuit, like at times the liver cells would simply be swollen due to viruses infecting the system either coming in through floppy drives, the feco-oral route or slipping in through emails, the daily sex routine that most Con's couldn't resist. Sometimes they would even seep in through infected needles that might have been jabbed into it in the past like software programs given with intent to treat but unfortunately themselves containing viruses due to faulty sterilization techniques. Once inside the system these viruses usually would manifest as acute system malfunction with a yucky appetite and yellow eyes for some time.

The systems gradually recovered as their immune software tackled the virus. However for some systems infected with real bad viruses like B or C the inbuilt immune software had a real tough time in eliminating the viruses. (For B it was difficult in ten out of hundred people inoculated and for C almost impossible for eighty out of hundred).

They may have been helped with artificial software like interferon alpha or Lamivudine, which at best eliminate the virus in only thirty to forty of hundred systems already inoculated. Most of the other systems un-helped by artificial software go on to develop what professor mechanic would label as a morphologic diagnosis of Cirrhosis of liver but here unlike Ram Hari's the etiology would be viral hepatitis either B or C.

Chapter 67

THE INOCULATED SYSTEM

Mr. Conwin Simon sat waiting his turn at Professor Mechanic's liver clinic with a worried expression displayed on his LCD screen. His blood tests suggested that his system was infected with a bad virus that was possibly fast damaging valuable files on his hard disc.

One of his friends had recommended Professor Mechanic's liver clinic and he was quite hopeful and yet couldn't stop worrying. Also he was feeling a peculiar heaviness in his lower limbs especially since last night after his wife noticed they were swollen. That was definite sign of a promising future system trouble, the very thought of which made him even more depressed. This was on top of the fact that his business hadn't been doing well lately.

The sweet shop he ran was affiliated to a popular brand, "Slurp", with a chain of monoclonal outlets around the countryside. Recently the company had suddenly decided to close down countrywide (although his particular shop was selling well). There was some story about the son in law taking over after the owner father in law's death and selling his company off for a better business nearer to his own heart. Conwin had no immediate options but sell sweets from local small time sweet manufacturers in the country.

His customers who relished the "Slurp" brand didn't take kindly to the fact that Conwin was still using the name "Slurp" in the neon sign outside his shop (originally provided by the company who didn't bother to reclaim it) while selling unbranded non media hyped local goods (which Conwin personally felt weren't less tasty). Conwin was angry at "Slurp" for having ditched him and other "Slurp" outlet owners who were making reasonable money to feed their families, angry at globalization for having given birth to "Slurp" in his small town business that was otherwise getting along fine without it before it came along with promises of slurp-more (slurp-plus) money.

His anger was soon quelled by fear, the fear of not finding another suitable 'slurp' or 'burp', the fear of resulting penury, eviction of his bright child from the local good school due to non-payment of fees, having to resort to selling his wife's jewelry and now this virus. He felt his system crashing along with his whole world coming to a hanged standstill. Professor mechanic informed him that the virus had already made sufficient forays into most areas inside his liver, systematically destroying each and every hepatocyte file. His leg swelling was due to an excess fluid in between the cells in his leg. The interstitial ocean showed up as edema due to an excess fluid of three to five liters inside his Earth. There was excess fluid also inside his belly, in the peritoneal bag covering his intestines along with other organs inside his abdomen.

This flood of fluid explained professor Mechanic, resulted mainly from a fibrosed liver with blocked inlet pipes because of repeated inflammation and healing around its microscopic hepatocytes all stimulated by an idiotic hitchhiking virus. Hep-B, the professor called it (a bit rather too affectionately thought Conwin). It also struck him odd that Professor Mechanic should say that the virus wasn't responsible in itself for the destruction of his hepatocytes. It was his own system's immune cells that were responsible for all the carnage as they went on a trigger-happy shooting spree burning off all the hepatocytes in their attempts to target Hep-B.

Chapter 68

THE LETTER FROM A VIRUS 1-5

VIRUS 1

June found Mr. Conwin very polite and co-operative soon after his admission to the ward, inspite of the fact that he was obviously depressed. On doing a physical examination she found his limbs swollen along with the fluid in his belly. This was removed by her senior for testing to confirm that the fluid was all due to an increased pressure in his liver. That night Mr. Conwin couldn't sleep well.

He woke up next morning to find June at her bedside. He was feeling groggy and she seemed to be asking him to keep his hands extended from his wrists for some time. He wondered why one would want him to do that. Nothing seemed to be making much sense any more since he had been admitted to the hospital. Last night he had received this letter from the virus that was supposed to be habiting his system, (which from the virus's microcosmic perspective) seemed a huge universe of its own.

Dear Conwin,

I am a Hep-B virus amongst the now countless many inside your system liver. It is a desperate attempt to reach out and communicate a tale of our life long struggle in a community hated and hunted from its very inception. The present state of affairs is such that we are nearing a perishable climax but more importantly, we seem to be dragging down your system the body of Earth as well.

We were originally a tribe of fishermen living off the ocean, in the west coast of a huge landmass of Hepatocytes. Our earliest settlers had arrived on its shores long back and a strange tale of their appearance has passed down verbally over generations. They claim to have frequented a similar planet, with oceans, islands, palm beaches and plenty of fun and frolic.

Once, on a yachting trip, his majesty, the HBV virion along with HBeAg and HbsAg, two of our ancestors (at that time probably in their teens) rode the waves a trifle bit longer than usual until the main land mass of Hepatocytes grew smaller and smaller, at a distance. There was a dark orange hue to the sky as the sun left an uneasy calm. It all began at that moment when both of them prepared to tackle a rather gigantic wave approaching and lifting them up. It became larger and larger, assuming proportions of a gigantic mountain.

Preparing to meet death, they suddenly found a huge steel tunnel piercing the sky and descending on to the ocean. They watched open mouthed as the gigantic tube sucked all the water along with them into its claustrophobic insides. It was, as if they were swept into another dimension and some of us wondered if it was a near death experience. After, a time, there was another gush of water from the other end of the tunnel, probably water being emptied somewhere, thought HBsAg as he hung on to the sides of the tunnel with all his might.

Before, long their tunnel pierced again into another good old three-dimensional earth and as soon as it touched the ocean, our boys started swimming for dear life. This time there was no suction vortex and they managed to swim over to a similar landmass of Hepatocytes. It was a complete virgin land in those days. None of the gory sky scrappers you see in the present concrete jungle. There were lush green fields bordering on dense forests, a hut or two, a fire to keep you warm for the night and that was all the smoke it produced. Not like the smokes of presently upcoming power plants.

VIRUS 2

Our ancestors arrived like Columbus and started knocking on the doors of the native population which resided peculiarly deep within the earth in a nucleus. Our ancestors might have opened a hidden trap door and descended down an escalator to a huge sphere with multiple small windows. Clothes weren't allowed and so his majesty, took leave of HBsAg and HBeAg, shed off his royal robes and climbed into the sphere, as naked as a DNA. There, he was greeted by the local chromatin population and shown around.

Great place! Thought the king and settled down, had children, another life, and another kingdom. Now we Hep -B viruses don't enter into sexual relations that easily but prefer to replicate our own DNA, so the king sat to meditate.

His deficiencies and gaps were filled and he became a covalently closed circular DNA (or you could say he was totally covered up in a termite mound). The CCC-DNA was extruded out of one of the windows and then the mound broke into three pieces of slithering snakes, ready for transcription. It proceeded with the synthesis of a minus strand followed by a plus strand and finally we had baby HBVs.

The first threat of an external aggressor appeared around this period. It wasn't anything to do with the natives within the nucleus who were pretty docile and never troubled us. It was the ocean predators.

They were probably the other inhabitants of this vast body of earth in areas, which still remain unexplored. They started off by attacking our children. Dear little baby peptides dressed in colorful epitopes loved playing in the beach. We never thought they would draw the attention of those detestable classes of MHCs. They would play with our children and hold them up in the air from time to time. Little did we know that this would be a signal for their CD-8 (seedy-ate) ally that is until the first bombardment.

The hepatocyte shores went up in flames. Fire spread rapidly, lighting up the usually dark jungle silhouettes. Across another corner, in a dim glow, one could barely discern other dark shapes getting down from odd looking boats, starting to drill the earth with perforin drillers and pouring poison into those drilled perforin channels .

The ravages we have survived along with our coveted land are a tale often repeated. I guess, every society has its persecutors, bombardments and yet life begins anew. Many of our brothers devised increasingly cunning ways of eluding our seedy attackers. It's been a topic of research among you humans (some of us refer to you as Demons/Gods).

Mechanisms of persistence of Hep-B viruses in chronic carriers, is a hot issue, the complete unraveling of which might push the final nail into our coffins. However, before that I seriously believe and wish to convey that your body of earth itself is in jeopardy.

VIRUS 3

Mechanisms of HBV Persistence

Some of our HBV brethren left the hepatocyte shores and swam with the strong oceanic currents to seek refuge in other greener pastures...kidneys, brain, testes, pancreas and the like. The seedy cells couldn't pursue them all the way to these places apparently dissuaded by micro-vascular anatomic barriers, which surprisingly are lacking in the Hepatic sinusoids.

Other than this, some of us tried to fox the MHCs by mutating the epitopes in our babies to evade recognition by seedy eyes. Then there were some peculiar natural factors, which came to our rescue as well. The MHCs we learnt were essentially polymorphic and all MHCs were not equally bad and some of them just played with our kids. The MHC II came perched on macrophages and these giant whales required a daily rubbing of Vitamin D oil on their surface, to actually stay active.

Recently some of your people discovered a mutation (T to C at 352 position) in the Vitamin D receptor gene that might lead to inadequate activation of these killer whales. These whales also love to eat with a particular sauce (mannose binding lectin) and again this was successfully adulterated.

Another ploy that we had to devise was born out of sheer desperation in our continued struggle against the seedy cells. One night, hordes of our guerilla forces attacked a fleet of passing PBMCs (peripheral blood mononuclear cells), climbed onto their ships and took over after a bloody battle. Their wireless system with which they used to summon more and more seedy cells were effectively silenced.

VIRUS 4

Carcinogenesis

Other avenues of survival opened up in our very homes inside the hepatocytes. The local people we noticed were surprisingly immune to seedy attackers. They had disagreeable features but in the interest of survival we started marrying into their communities (effectively, integrating with our host genomes). We married and multiplied into a modern cosmopolitan society. Gradually, we lost our food gathering nomadic existence and settled into concrete stable dwellings with equally fine-processed food to satisfy us. Multi-storied buildings started adorning our skylines. Gone were the palm trees, green fields were encroached by collagenous cement that hardened with each passing day. You could almost hear the muffled cries of earth as it stifled over this new overpowering burden.

Since the early part of this century ...we had continued to rebuild and regenerate after each and every seedy attack. However regeneration and degeneration can leave ugly scars. Presently we have had the unfolding of another interesting scenario. The seedy missiles having been silenced, what we have now is simply growth and more and more growth.

Some of us have begun to see through all this growth and realize our days as well as that of the earth are numbered. It's only a matter of months now. That is, if we can't devise new ingenious methods to escape our present predicament. Life is generally defined by some universal traits, which we suspect might be present even in you humans. It might be present even in the myriad subatomic particles that make up our relatively puny bodies.

VIRUS 5

The earth which we live in seems similar to your earth and we are accumulating enough evidence that it's inside you humans where we might be residing (like your concept of a living earth). Therefore, this uncontrolled growth of our civilization might be the death of your body and therefore this desperate communication (By the way, is the growth of your society too progressing exponentially - really, this concept of parallel universes is fascinating).

Life grows as an expression of its universal traits, chief among which is a perennial dissatisfaction with the present and an insatiable urge to change, replicate and evolve. Look at our evolved bodies, added up from myriads of subatomic particles and look at your body (our earth) even more highly evolved and yet yearning for more. More love, more growth, more reproduction both in thought and form ... yet survival is the name of the game and we have evolved over the years, just to survive. I am confident some of us will manage it even in the wake of this global cancer.

As night falls, a few dark silhouettes among us have been noticed to wade out with their rubber dinghies into the deeper waters of a darker ocean, hoping against hope.... someday ... another tunnel, another universe, fresh green rice fields bordering dense forests. Clouds you'll notice, too exhibit living characteristics of a non-linear replication, summation and accumulation, which from time to time, bursts into a downpour, catching you unawares in the rain. We wanted to live like a cloud but ended blowing up in smoke.

Yours truly,
Hep-B alpha

Conwin had suffered a blood vomit that very night of his first day of admission and in the morning he was found by June to be a bit drowsy. He was in a confused state and June found his hands flapping when he tried to obey her request for keeping his hands extended (a phenomenon that Professor Mechanic for some reason liked to call asterixis and believed it signified a diffuse problem in all the cerebral neurons which again was in all likelihood due to chemical-metabolic and not structural reasons).

An ultrasound examination that very evening he got admitted had shown a tumor in his liver. Mr. Conwin died the following day. "Mercifully saved," June thought from all the treatment that would have followed if he had survived. Conwin's elder brother had come to take his corpse and on his request she removed some of the fluid from his lifeless belly as that definitely made it lighter and easier for them to carry.

Chapter 69

HUNGER FOR FOOD AND AIR

June now in her 7th semester is posted in the tuberculosis ward of her hospital right in the heart of the city of Kolkata, a heterogeneous mix of complex systems animated and inanimate alike. She gets to meet a lot of patients every day adding to her narrative database now rich with innumerable patient anecdotes. She sits at the bedside of a lady and tries to start a conversation by asking the usual what is your problem and gets a well-framed morphologic and etiologic diagnosis as an answer, instead of the usual symptoms often narrated by patients.

The lady clarifies that she is a physician, a former student of this college and now admitted as she has developed tuberculosis on top of her newly diagnosed diabetes. As she talks with June her mobile phone rings and June gets a feeling from her telephonic conversation that she must be a successful general practitioner. "It's quite trying you know to remain in a hospital bed while you have a busy patient practice." June wished she could be like her in the near future.

She continued her patient interviews and was taken up by another lady who appeared pale and yet cheerful as June approached her. She had been diagnosed with tuberculosis long back 10 years ago she said when she was just 25, that was a few months after her marriage and her husband had left her in the village to come to this city in search of a job and never returned since. She had lived off her parents and brother for a while but had to leave them too eventually, as they couldn't bear the cost of her treatment. She took up the job of a sweeper in the local health center in the next village. She learnt quite a few medical things there as the doctors and paramedics in the center would use her as an all purpose help and village patients soon became used to her sweeping the floor of the clinic and at times dropping the broom for the hypodermic syringe that she would stick into their posteriors with equal élan.

On doing a physical June discovered she had an increased breathing rate more than normal and there was a peculiar caving in of her right side of upper chest. The sounds too of her breathing that she heard in that area of her chest were definitely not normal. They were more of hollow, tubular rather than the normal soft rustling. Even tapping on that area with her fingers produced a dull sound rather than a resonant note. She could also see that the big air pipe into her lungs and her palpable heartbeat had shifted to the same side as the caved in area of her chest. She asked the other lady doctor in the room about what morphologic diagnosis that would suggest and she said, "Looks like a right upper lobe of lung collapse possibly due to fibrosis as a result of her old tuberculosis."

As June got ready to leave, the patient suddenly asked her if she could request the hospital authorities to ask the kitchen help to give her a little more rice in her lunch as she felt so very hungry even after finishing off whatever they provide. Looking at the surprised expression on June's face the lady doctor laughed and said, "These poor patients never get enough to eat. The food they give them is horrible and yet they hunger for more. Actually the government has sanctioned enough money for good and adequate food for them at each meal but half of it unfortunately makes its way into the neighboring restaurants sold off by the hospital kitchen staff who have a shady liaison with them. This is our country, as physicians we can only practice and make money but we can't have our own sincere research. We leave the research to the West, as we don't have enough to eat."

Chapter 70

AIRWAYS OPENING INTO SYSTEM-RIVERS

Professor Mechanic resumed, "Approaching patients with disease in the lung and the bags covering it can be tricky as well as easy at times. The lungs can collapse like punctured balloons either due to an obstruction in its air pipes, the air in it getting absorbed by its rivers or they could cave in due to fibrosis that follows any inflammation. The balloon like air spaces inside the lungs can be filled with pus due to infection or they may develop tumors.

I shall take you to the Lung wards where we can meet an assembly line of systems with afflicted lungs in all its variety. You see this 35-year-old lady she has a caved in right upper chest with a dull note on finger tapping and a breathing sound that is hollow tubular...Collapsed right upper lobe due to old tubercular fibrosis. This 25-year-old girl with wheezy sounds from her chest and breathing difficulty has a spasm in her airways because of an inflammation there again stimulated by her system immune cells responding to foreign particles. This 50-year-old smoker has just the opposite of what you saw in the 35-year-old lady. He has a bulge in his right upper chest instead of a caving in but then the dull note on finger tapping and the hollow tubular sound are also there. This patient has a tumor within (what we call a morphologic diagnosis of right upper-lobe lung mass).

Post modern post bed-side student conversation

Tap-tap the back of this patient on his right, a golden backed woodpecker hammering on an old oak tree...Sounds normal eh! Tap-tap the front of his chest...um...sounds dull on the right upper part, the distant thud of a woodcutter at work, deep inside a dense forest. Now place your stethoscope on its back, the gentle rustle of a summer breeze traveling through pine forests bordering on moonlit sands. A normal vesicular breath sound right!

Now listen to the dull area you just tap-tapped. Uh! Oh! A bunch of jungle babblers engaged in furious conversation. What does it look like? Um...fine discontinuous high-pitched crackling sounds. More like firewood sputtering eh! Possibilities are, right upper lobe fibrosis due to old tuberculosis or right upper lobe collapse due to an endo-bronchial growth with post obstructive pneumonia.

Wait a minute! What's that growling sound?

Aw! Gwan...I remember the fright you gave me in Ranthambore! It turned out to be only just an insect and not a tiger after all. I am sure...I'll be...there's definitely something in there...deep down...maybe. Oh well, we do know that lesions deeper than 5 cm within the lung may not be audible. Could it be a solitary pulmonary nodule? Never heard of it producing such a sound, a solitary hermit perhaps deep inside a forest, engaged in conversation with a tiger.

Chapter 71

MOVING MUSCLES AND STORIES

June was waiting in the hospital corridor for her case to be allotted when she suddenly noticed an interesting machine walking in a peculiar manner with his limbs snipping the air between his thighs like a pair of delicate scissors. He didn't seem to be in much of an agony but did have difficulty in dragging his stiff limbs. June knew this had to be an electrical problem and eagerly followed him to his bed.

"Machine movement, began Prof Mechanic, is primarily due to muscle cells attached like flexible bands over the rigid skeletal rods termed system bones and with each and every electrical stimuli there is a contraction of all these muscle cells that make the flexible bands contract producing a torque to the skeletal rods, resulting in movement. Movement could be impaired due to either the muscles drying up on their own fashionably termed myopathy or due to improper electrical firing into the muscles. If the current flows through them too much it causes the muscles to go awry resulting in stiffness of the limbs.

This too much of erratic current could be due to a loss of inhibitory signals from the ICs in the spinal generators because of damage to their wires (fashionably termed the upper motor neuron, which indicates their source and distribution). The wires of the upper motor neuron start from cells in the cerebral cortices, which most machines carry in a skull casing that also houses its own LCD display at the front, termed face, replete with more muscles and finer circuits for facial expression. The same wires travel through an internal capsular area and the brain stem inside the skull, in pyramidal bundles and finally reach a station labeled "Anterior horn" where the upper motor neuron train stops. Another bunch of neuron wires called the 'lower motor ' take off from there and reach all the muscles in the human machine.

As current increases in upper motor wire damage it leads to a chaotic state of increased muscle contractility making them spastic. Lower motor wire damage has the opposite effect and makes them flaccid and limp. The machine is hardly able to move a muscle at all. A galvanometer testing the muscles would register no current. With time the muscles atrophy and the limb is just a rod."

Chapter 72

POSSIBILITIES JUMPING AND QUIVERING

June followed the young man to his bed with her galvanometer which was this plastic thing resembling a hammer. He was a bright boy preparing for his pre-meds and had topped his class in his village schools up in the mountains of Garhwal.

His muscles didn't look wasted at all. His muscle power wasn't that bad either, in fact it nearly equaled June's own as she asked him to flex his elbows but then on striking his muscle tendons in his upper and lower limbs June found them jumping and quivering.

It was evident that the current flowing through them was quite strong but obviously misdirected due to malfunctioning ICs in his spinal generators. Somewhere in the thick bunch of wires especially as they ran in the area of his spinal cord at the back of his neck (termed the 4th cervical segment) was the lesion (the morphologic diagnosis). What was it that was disrupting the ICs there?

Was it a tumor? (Unlikely...the machine had almost grown up with the malfunction). Was it the skeletal vertebral casing compressing the ICs causing cord dysfunction that was something treatable? Again looking at his back June couldn't find anything wrong with his vertebral column. Was it some blood pipe to the spinal circuit that had got blocked, drying up the ICs? Was it an inflammatory demyelination, a condition where the ICs lose their myelin sheaths (they help the current flow, jumping saltatorily over their nodes of Ranvier)? Unlikely again because both would have had an acute onset and the latter would have shown signs of recovery as soon as a new sheath replaced the old one in the ICs of the cord.

It looked more like as if the cells forming the ICs in the spinal cord had died on their own, a process termed apoptosis, which might have been triggered off by an external event, the entry of an email virus or may have been a consequence of a faulty genetic software program inbuilt into his operating system.

Chapter 73

SEEMS AS IF IT WAS JUST THE OTHER DAY

Kris noticed June for the first time in Samsara's *akhara* during the winter festival in Bolpur, a place enriched by the past of India's poet laureate, a university not very far from Samsara's own at Sushunia. Kris had simply gone to Samsara's *akhara* out of curiousity. He wondered how a professor of anthropology masquerading as a Baul would look in person; however it was the professor's daughter who caught his eye and tugged at his heart.

Juneli Sen he knew had topped that year's primary school living certificate exams among numerous students who had appeared. Krishnadas Mitra hadn't done too badly for himself but topping the list was like a wild dream. Everyone had complimented Samsara adding that his daughter had inherited her father's brains. In Samsara's hard disc inoculated with Baul spirituality there were areas that still responded to ego titillations.

Kris spotted June with her father singing the Baul songs he loved. She had a strong voice that seemed to have been well honed in Indian classical. They were flanked by Samsara's students who were also eyeing June. Kris didn't know how to approach her. She had this long hair in plaited braids and a smile that melted Kris. He wanted badly to know more about her, talk to her and then suddenly it all just happened. Samsara had begun singing and as the audience reshuffled he found June squeezing in beside her both of them sitting cross legged on the floor with one of June's knees snugly nested against Kris's thigh, both hidden behind a thick layer of Sari and trousers.

It was June who spoke first. "Are you an Engineering student?" Kris gulped (only momentarily), "Well no. I am Krishna, my friends call me Kris. I have heard about you and it's a great pleasure meeting you this close."

As June and Chris walked into the moonlit night they crossed the *Saal Bithi* , a small forest of Sal(Shorea Robusta). June had particularly requested Kris to accompany her through this forest to her hostel gate, as the forest didn't just have trees. The trees watched them, protector and protected as they walked side by side, their hands just brushing each other, as if not daring to do anything more.

Con had never experienced anything like this before with his machine. Love was in the air knocking at the doors of Consciousness (Con for short). What was it in the physical appearances of their machines that attracted them? Was it the way June smiled, her dimples creating a bewitching exterior? Was it in the manner Kris carried himself, the dreamy look in his eyes promising a great future (that was at present hidden in his side bag manifesto). They talked of all sorts of things under the stars and the trees and knew not when time flew or hopped from tree to tree just to cover that short distance from the fair grounds to June's hostel.

Chapter 74

LIKE A WHIFF OF RAIN SWEPT EARTH

After the secondary school leaving certificate, which June topped there was this higher secondary school leaving certificate that needed to be topped and Kris was healthy competition. Both would meet more often exchanging notes, differential calculi and physiology of the cockroach amidst a few unconventional write ups that June would pen down from time to time in her diary, again a legacy inherited from her illustrious father.

June's Diary

Dear Kris,

Sometimes, it seems as if it was just the other day you turned up suddenly out of the blue, gave us a pleasant surprise. Remember…the roller coaster ride on top of a rainbow, stopping to chat with a flock of geese returning, sleeping awhile on soft cumulus clouds.

You've been away for long yet, now that its time for sunset, the stars seem to lie in wait for someone, before they start peering out, in countless dazzling millions. If you could manage some time do visit us soon. Long since we bathed together in the darkness and quenched our thirst on the other side of the moon.

Kris's Note:

Dear June, You stepped into my dream, like a whiff of rain swept Earth. A cool breeze blew through the windows while the rain…created music outside. The room bore your silent eyes and me lying on the mat. From an unearthly distance you bent down and breathed softly into me, while I …kept on struggling to reach you, through dense jungles, steep slopes, gradually you receded into that never ending corridor.

The results were out and June had topped again this time securing admissions into most of the prestigious Medical and Engineering colleges in India. Kris hadn't done badly for himself either and had been lucky enough to get himself a seat in a medical college in Kolkata. June left her seat in the prestigious central institute for Trouble shooting humans in Delhi just to be with Kris. The distance from Sushunia to Delhi wouldn't be cost effective if she had to come and visit them once a month, she told Samsara. Samsara was anyway overjoyed and heart broken at the same time. Joyful for his daughter inheriting his brains and saddened that she was joining the very subject he loathed. However he had decided he would let her make the choice rather than decide for her as his parents had done for him decades back and ruined his career. “I thought my daughter would have made an intelligent choice but no she had to become a doctor knowing very well the soup I had landed myself in.” Maya comforted him, “It’s possible that she wants to avenge your failures in medicine.”

Chapter 75

EEEEEK, OR SHOULD IT BE MMMMMMMMMMMF...I SNIFF SOME PAIN, A DEEP BREATH OR TWO

June and Kris pass their medicine finals and June is expectedly at the top of the list. Both start their internship enthusiastically. However, June notices Kris gradually starting to ignore her. One day Kris discloses his plans to go to US for which he's being financed by a mysterious uncle. June can't afford it herself. Samsara with his meager professor's salary can hardly cater to all the demands of Maya and Dupki, her younger brother. Also money would be needed for her brother's engineering school and Samsara's Baulish antics were far from paying. June would have just loved to go to US. If only Kris would have offered to look after her there, but no he was being very secretive about his uncle in US who wanted to see him there alone.

Most of June's class acknowledged that even if one single person was to score the highest in the US medical licentiate exams it would be June, given her marvelous track record. June had to stay back and see off Kris at the airport with tears in her eyes and she had a feeling she would be seeing him for the last time. One of her friends generously volunteered the information that Kris's mysterious uncle sponsoring him in US was actually his would be father in law.

The Indian Institute of Trouble-shooting-humans(IIT), post graduate entrance exams were the toughest in the country not only because it was an institute of national importance with all Governmental grants directed towards it, that made it highly covetable but also because of a system of in house reservation for students who had already graduated from that institute. Off course they had only been able to do their graduation from there because they had cleared another tough entrance exam at the pre med school level, the one that June too had cleared but had forfeited in preference to studying at medical college, Kolkata, just to be with Chris.

Now that Chris was out of her life, June was free to try out her luck again this time at the IIT for gaining entry at a post graduate level. June never needed much luck. Or is it that she was lucky to have got such marvelous brains. A photographic memory allowed her effortless data storage in her hard-disc with an infinite capacity. In addition a super fast clock speed with a very high random access memory allowed her effortless data retrieval. If life had been hard to June in matters of heart it was extra kind to her brain. She was relieved to see her name in the merit list right at number one. Relieved not only because she had something good to fall back on (if not US) but she was also going to get a good stipend as a post graduate resident. This would even allow her to sponsor her brother's undergraduate engineering school.

Chapter 76

HOUSE PLANTS CRUISING THROUGH SUMMERS BLISS

June decided it was time for her to chill out and relax on the system trouble-shooting after all, things were not that dim for the future. She had a secure job as an MD resident and there was a saying about the institute that if one just managed to keep one's eyes and ears open while working in the hardware lab located in the wards of this teaching hospital, knowledge would seep in from its walls. These walls would nevertheless confine her in their midst from 7 AM to 10 PM. It would also float in the fresh cool air in the institute gardens (on rare free Sundays) or the hot air in its canteens (where heated lunch and dinner time discussions would manage to keep the food palatable).

She met Seshu in one of those rare picnics that had been arranged by the Gynecology department. The bus took them to a place near Delhi called Barog, a beautiful hamlet near the Himalayan foothills. Seshu was quite lively, singing and dancing whenever given the opportunity in impromptu picnic sessions. June found an empty chair beside Seshu at dinner time in the bungalow like hotel and Seshu immediately struck up a conversation. Other residents eyed them knowingly as June appeared to be getting quite taken in by Seshu's advances (the picnic was a success even if it just managed to pair up a single couple). They were two separate departments and June could meet him only during meal times in the canteen.

Seshadri Srinivasan looked jubilant on meeting her each day and they would talk on a variety of topics under the sun. June liked him but...there were too many buts and the important ones off-putting about Seshu was that he was by caste a Tam Brahm (even if he wanted his parents wouldn't agree) also he was much too fast and ethereal. Krishna's wounds were yet to heal, although Seshu did provide the much needed balm. Predictably Seshu proposed (a steady relationship) by the end of the month and was sweetly turned down by June to which his predictable response again was, "We could still be friends." Far from it he quickly started chasing other girls in the campus. He had once confessed to one of their common friends that having a good time with girls was like a train he was afraid he might miss if he was too late...time was running out for him and so was life. He had to part ways with June just to be on the move.

June let the rigors of residency wrap her up like a cloud covering a blue mountain and bathed daily in the interesting discussions that would take place over the trolley carrying patient files, ivy sets and spirit cotton. Most of the machines in this high brow institute would have been admitted for a major dysfunction, many a times life threatening (except the odd one or two admissions catering to relatives of political leaders who kept needling the consultants with their phone calls). June for the first time was playing an active role in system trouble shooting. She was a registered mechanic now, had the license to prescribe, input software programs that could eliminate system bacteria, viruses, open up blocked pipes, flush out water from the kidneys and very importantly be able to suppress the system's immune system gone awry. She had the power and ability to decide which system required what program (at least make urgent decisions at her level when her seniors or consultants weren't around).

Chapter 77

PIERCING INDIVIDUAL UNIVERSES

Things now were quite different from her undergraduate days, where according to the system prevailing all she did was read books and observe machines and their afflictions evolving, hardly ever needing to take responsibility for any of them. In MD residency she was responsible for the management of each of her allotted machines, beginning with daily drawing samples for analyzing system river data that would in turn reflect the functioning of each individual system component. She also had to fix input-output (I/O) devices on to the body of her systems allowing access to their rivers within.

There were commercially made small and large colorful IV cannulas that helped input drugs and fluids into the systems along with gauging their central venous pressures. The latter was a larger cannula inserted in bigger veins draining nearer to the central pump, doing which was technically challenging and involved trying to scale a steep reflex learning curve. Apart from assessing central venous pressures, the machines had a lot more outputs to be analyzed like all the chemicals in its fluid firmament dazzling brightly like molecules of countless stars.

The only way to really know the chemical compositions of these star constituents was to delve into system-rivers, sampling them every morning with painful needles sucking in blood and sending them to the lab in buckets full of test tubes. The laboratory physician in his observatory would take readings throughout the day from his spectroscope. These readings were from reflected light that bounce back from the stars, blood molecules collected from pierced individual universes. The residents would pounce on the evening reports as soon as they arrived neatly attaching them on to their individual system file folders ready for the evening senior residents rounds. June ran her eye quickly over the reports of bed number 16. Sodium-140Meq, potassium-4.0Meq, Creatinine-1.2mg/dl, albumin-4 g/dl...no major biochemical problems in this individual's universe, the stars winked, another bed, another universe perhaps.

Then there was sampling that involved not just collecting water from placid normal rivers, venous tributaries but also areas where there had been floods. Excess fluid in the pleural cavity (a membranous bag covering the lungs and other pipes inside the thorax) would necessitate a needle jab at the back of the chest. Similarly fluid in the peritoneal cavity (the same membranous bag covering the intestines, liver, kidneys etc) would need a poke in the abdomen. The rivers around the spinal cord generators could be accessed with the same needle devices requiring a bit more technical competence that June picked up easily, having to practice the needling art in her daily voluminous sampling efforts along with her colleagues. Some of them boasted of the ability to tap the spinal fluid by throwing needles like darts from a distance aimed at the space between the 3rd and 4rth lumbar spines on their patient's backs.

Then there were needles that could sample bits of tissues from the individual's system components. Needles biting off pieces from the liver, kidneys, pleura, peritoneum, lymph nodes, gave a valuable peep into the microcosmic battles in those individual component organs. In between her ward duties she would also be posted in the casualty where her technical skills in establishing I/O access devices would be honed further. She even learnt to effortlessly intubate machines gasping for breath, thus helping to establish a vital airway connecting patient system pipes to external mechanical lung pumping systems.

Chapter 78

WAITING FOR THE MOUNTAIN

June made friends easily, only some of them were life-long. There was Rita who shared her interest in poetry, birds and microcosmic system architecture. Their first weekend trip was to Solan in Himachal where June saw mountains unlike the rock outcrops she was used to in Sushunia. Rita told her about even loftier snow clad ones visible from Shimla, a few more miles up and how they gradually started getting bigger and bigger as you went towards them further ahead to Rikong Piyo, Kalpa in Kinnaur. June had dreamed of them earlier during her school days on listening to anecdotes at rock climbing excursions from Samsara and his climbing friends who made yearly forays into those peaks. They were average Bengali lower middle class tea cup intellectuals working in Govt offices in Kolkata but once in the mountains they became transformed into climbers, industrious and efficient in laying fixed ropes across difficult long patches.

June's Diary, Sept. 01

There are some people who never get to see a mountain. All they can do is simply wait. Any of these days it might just happen, flapping its wings out of the blue would appear a shining desolate golden mountain and lift them off their park side benches.

12/9/01,

Female medical ward-1) Traumatic quadriparesis—Biceps brisk with fasciculations . CT Epidural hematoma compressing thecal sac. Power improved to gr 2.

ward1, Kamla's mother – pyrexia of unknown origin (PUO)-sent off by me from the out patient department (OPD) as viral…still intermittent fever spikes.

2)12/9/01,6:30PM-Abha Sharma 22F-Fever with hypotension-admitted yesterday -CXR-showed Lt lower zone consolidation...I added Crystalline penicillin…patient suddenly threw a seizure after having got crystalline penicillin although she was all right…just a bit restless after the attack. It was her attendants who were acting more hysterical and I had to shout at them and one point of time while I was looking for a vein to give her diazepam and her husband was muttering under his breath that we were continuing to experiment all sort of medicines on her I just blew my top. I coolly kept the injection syringe back in the table and said I won't treat her and asked her to take her wherever he wished. That changed him completely to the reverse and he started to apologize profusely. After that we shifted her to the ICU and put in a CVP line for her hypotension.

> The day one comes face to face with a huge mountain, automatically windows open, deep down somewhere inside the brain. Dark brown clouds atop a black massif penetrate your innermost corners and one fine day they suddenly emerge barging into a busy office room full of overworked colleagues.
>
> June

Chapter 79

FLYING IN A MAGIC CARPET

Residency days were flying past quickly. June was having a great time.

June's Diary

Ward notes Oct 21,2001

ICU-

1) Kataria---Fever persistant ?aspergillosis ...ampho available --prob--hypotension-fluids not lifting bp-Died in the evening
2) Shanta Singh76735- 91M

left ventricular failure-better—Chest X ray (CXR)-Pulmonary Edema-surprise! echo - only left ventricular hypertrophy (LVH) with normal systolic function!!
22-10-01-Last night -Restless again –6-11-01-Came back with complete heart block due to digoxin toxicity-accidental administration of four tabs of digoxin by son thinking four tabs would add up to 0.125 mg as had been advised.

10:00 PM Came back with Rita on a rickshaw after seeing a Bollywood movie. The rickshaw puller allowed me to drive it myself on the deserted night roads much to Rita's chagrin.

I strongly believe amidst us somewhere
Resides a gigantic mountain
Full of perennial snow
Shining brightly in the afternoon sun
Changing colors with time
Morning…a pinkish white
Evening…a orange hue
We go about all our lives
Trying to scale our individual mountains
Somewhere deep within them
We exist…ourselves

Diary Entry made in a Moving Train

A two-day respite, from a busy workplace… Retreat…to a bird-lovers paradise…A small and sleepy hamlet in Rajasthan to which birds of all kinds of feathers flock. An intimidating distance from our Institute and yet if our feathered friends can take all the trouble to visit the place from more impressive distances, why can't we…these are some of the thoughts which tend to comfort me while I tremble and freeze inside the Chattisgarh express as it tears its way through a chilly Uttar Pradesh country side in the middle of a winter night.

Chapter 80

BIRD BILLIONS, BEEPING TWEETING WHISTLING WITH THE RIVER-TUNES

June's Diary

Next day I get down at Mathura. Bharatpur's another hour ride by bus and after keeping my belongings at a relative's place I take off in eager anticipation to meet the birds. "To ghana ..."I order the rickshaw wallah who obviously takes me for a local as it'd be unexpected for a tourist to know the name by which local people call the sanctuary. At the gate one has to deposit his/her identity card and Rs 20/- in exchange for a bicycle and I set off merrily through the road cutting across the jungle.

Birds

On the way you meet the usual fare of red vented bulbuls, jungle babblers, robin magpies, pied mynahs, golden backed woodpeckers, spotted owlets etc...which are commonly strewn around any Indian countryside. You also meet the wag- tails, white whiskered bulbuls, bay-backed shrikes and brahminy mynahs which to your surprise and delight seem to thrive here in greater numbers than anywhere else. As you pedal along on the rusty old bicycle you are suddenly confronted by an amazing and dazzling spectacle. Both to your right and left there seems to be a stage set on the marshlands in between which great big trees arise and atop them are a host of painted storks. Scores and thousands of them painting the whole of Bharatpur red with their beautiful red heads and necks by which one can make out they are adults. Their young ones create a nerve wrecking cacophony with their relentless string of chirpings and this only heightens the excitement of the avid bird watcher as he takes up on the challenge of unfurling further the mysteries of this fascinating patch of land. Apart from these South Indian friends we also have at the same stage the black necked stork on whom you may chance upon suddenly and get a shock as you behold a grand looking bird in a peculiar posture of anticipatory attack, still and waiting...probably for fish. Also we have, snake birds and cormorants in the dozen who have a peculiar habit of spreading out their wings to dry in the sun after each and every dive probably due to a lack of oil glands in their body unlike other birds that are more fortunate. You also see the grey and purple heron as well as the purple moorhens on almost every patch of marshy land you lay your eyes on. The moorhens are strikingly purple with a bright red crest on top of their heads just like hens and the purple color of the heron can be made out only when they prepare for flight.

People

The people here can be immediately classified into, the already initiated and expert bird watchers with their slick binoculars and zoom lens and occasional references to Collins or Salim Ali's and the uninitiated whom you can spot on the rickshaws…mumbling and memorizing the awful garble of bird names"….purple…purple…what was that he said!"…Moorhen! Really dad! Why do you have to be so slow…oh! It's just the age my girl!"…Amazing rickshaw-pullers rattle off bird names and descriptions smoothly and rapidly much to the consternation of our newly inquisitive bird enthusiasts who have to really struggle to keep pace with them.

Chapter 81

THE BIRDING RESIDENT

June's Diary

As you stroll along you are suddenly confronted by a sign "would you like to be born as an endangered species in your next rebirth...if not please help save the Siberian crane..."its then that I suddenly have a chance to meet the star attraction of Bharatpur. A kindly man lends me his binoculars to have a better look. A beautiful reddish beak with a large white body and a peculiar tail, as if somebody has combed his hair forward. There it was pecking at something, in all its majesty...I wonder if she knew her worth.

Countryside

Now for a boat ride...you are transported through dense dark forests surrounding you on all sides and at times you get to see Indian moorhens scuttling off as your boat approaches. You are brought very near the nests of the painted storks where you can see the young ones screaming for food. You can also marvel at the open billed storks you had missed earlier on your cycle ride. As the sun sets in Bharatpur you are treated to a wondrous spectacle of an orange glow emanating from the swamp waters with the dark silhouettes of the herons and storks atop the acacia trees creating an everlasting impression: a picture that would arise from the deeper most circuits of your cerebral networks on a midsummer day, in between a busy and tight schedule of commitments and conferences, a silent and comforting oasis in the midst of an arid desert of city life.

June returned to the rigors of residency and the approaching exams for completion of her MD degree. The dissertation submission had in itself been tough as it had involved a lot of last minute changes before the dead line but once finished June had set out on this birding holiday. There were a lot of things to read for the exams. Trouble shooting humans involved having an extensive knowledge of the present technicalities, stories of the microcosm that have evolved over generations, more so exponentially in the last few decades.

Chapter 82

THE BRONCHIAL TREE

Take a look at this chest x-ray...Hmm dense forests bordering on green rice fields

A cow munching grass, lazily waving away the flies, the setting sun, and one or two clouds interspersed ...Hmm looks normal-show the other one. What's this! A giant jungle mower, Concrete new buildings-all cooped up look alike? Hey can you hear the trees crashing!

Can't make out – over here the mower's creating a din. Hmm see anything more?

A lot of alveolar opacities which is probably the smoke from our upcoming power plant...

Hmm the diagnosis is clear—Lung cancer. Get a bronchoscopy but the prognosis is bleak

At the most 3-6 months.

Sutra

He was known as Dada to his friends. He had remarkably long hair for an Indian along with a long beard. They say he used to religiously tie it in a turban once everyday before they attacked him mercilessly one day when the prime minister died, killed by another person with a turban. That day there were countless such victims in the roads of Kolkata. It's a city where outdated Government tram-cars are usually torched by violent mobs at even lesser pretexts (the price hike in the tram ticket being one such routine stimulus). Dada was transformed that day not only because of the hideous gash inflicted on his face by a religious fanatic but strangely the whole experience made him very unkemptly irreligious in the years to come. He let his hair down like a hippie and started smoking like a chimney (smoking was strictly forbidden in his religion). Another development that was totally unlike Dada in his pre-hippie days was his fascination for poetry. He was a permanent fixture on the Presidency college campus spouting a free flowing stream of consciousness verse that would intimidate even the utmost post-post modernists.

June came to know him only as a solemn retired professor of chemistry from Punjabi university, introduced to her by the senior resident as one with interesting findings on his chest, a respiratory case that would be surely kept for her ensuing clinical exams if available at that time. Sure enough there was a bulge at the right upper part of his chest that was dull on tapping and there was a hollow sound on putting the steth. A mass lesion is the most likely morphologic diagnosis thought June. She had become quite deft in her handling of cases during her residency years.

She chatted with her patient as always but couldn't find time to read what Amrinder Singh 'Dada' had written in his diary.

> Last night I discovered the fruits of bronchogenic carcinoma growing on my bronchial tree.
>
> Ugly and slimy it tasted of blood and phlegm. This tree…had long been a source of shade for travel weary souls amidst sun burnt fields. Not any more…falling leaves, shriveled bark, ugly nodular fruits with a slimy sauce. With each and every cough it shakes from its very roots caught in a raging storm.

Chapter 83

EVALUATING MECHANICAL TROUBLE SHOOTING SKILLS

The exam was an elaborate affair. There were 4 machines allotted to each examinee for each candidate's assessment of trouble shooting skills. These were labeled as long cases possibly because of the longer time allotted for their detailed evaluation by the candidate (as well as the longer time taken by the examiner to grill each candidate on the cases). They also carried more scoring points that decided the fate of the examinee. There were a few short cases with lesser marks allotted that also involved lesser discussion time.

June received the usual standard fare of system problems, a young man with abdominal swelling, which June correctly diagnosed clinically as due to a blockage in the outflow of the pipes from the liver particularly as this man had prominent veins in his back (that was an important clue to his diagnosis). The examiner looked unimpressed and insisted it might still be a problem with the liver per se and may not really have anything to do with the liver outflow pipes.

The second patient was a failed river pump due to a problem with its valves. June thought one of the valves was obstructed and the other leaking. Her examiner thought both were obstructed. It was all a matter of interpreting those whooshing rumbling sounds upon which the future of the examinees (if not the patients) depended. The third patient was a straight forward fibrosis of the lung and June localized it well right down to the lobe.

The last patient was the one that flunked June. It was an electrical circuit problem and June found it very difficult to get any kind of localization as the story offered by the patient was very sketchy and he seemed decidedly uncooperative and even unable to follow June. She made a diagnosis of a lesion in the spinal cord particularly its ICs at the level of L1-L2 (first and second lumber vertebra).

The examiner who was a lady grew wild with anger and asked June to examine the jaw jerk. To her surprise and dismay June found it was exaggerated, a vital clue that she had missed that would definitely suggest a lesion much above the ICs of the spinal generator, (above the mid pons in the brain stem to be precise). June finally realized why she couldn't get a good history from the patient; it was because he was demented. The examiner also showed her a few more signs that suggested that the localization was in the frontal lobe of the brain. June's fate was sealed. She would have to reappear again after 6 months.

Chapter 84

ON FAILING THE ACID TEST

June's Diary

On failure:

Three years in a superb academic institute with an exciting and rigorous curriculum. At the end of it a costly slip commands you to take the same exam again 6 months later armed with more knowledge. You always presumed this happened to other people and couldn't affect you but one day it does...and for that moment...it seems to be the end of the world as you go through the often quoted stages of denial, grief and finally...acceptance.

As the stars peep out into a desolate night you wonder...end of the world? It's just another world amongst all those millions of shining beauties. Amongst them, reside many more countless organisms, giving birth, struggling, dying to keep pace with their species...and in one such world...somebody just doesn't make it to the finishing line. It's as if you're observing yourself from a spaceship, a tiny speck under a satellite image, racing amongst the multitude.

However as days go by, you become aware of a new status...unemployed, minus accommodation, greeted with avoiding looks. Can't help it…for 3 years the ride had been smooth...might as well savor the bumps for another six months. Once you get used to it, it isn't as bad as one would expect. There's loads of free time for contemplation as you watch your co-residents rush to work struggling with their new found burden of responsibilities, the vagaries of senior residency. You can afford to sit back and relax, share a biscuit with the baby monkey scratching your door in the doctor's hostel.

You can go for long walks or read undisturbed in one corner of the library. Failure can mold a whole personality...its as if you're on an obstacle course and suddenly in the midst of a long run you discover a deep chasm impossible to cross. By the time you collect yourself, you find the rest of the herd has disappeared leaving a dust trail. You've to wait a long six months for another herd. As days go by, you press your ears on the ground straining to hear the sound of distant hooves. This time you have to make it. Each night you dream of the ever widening chasm, the unpredictable stumble over hidden stones. You must jump proper this time. Nothing should be left to chance.

Chapter 85

SEARCHING...SEARCHING YELLOW EYES

Con picks up another fragment, a yellow one like an evening sun ready to set. He sees himself as Juneli, Samsara's daughter who he lovingly calls June. June has grown up into a young studious lady with plaited braids and thick glasses. In this visual before you, she's a resident fellow in Hepatology and today's her on call day.

June's Diary

This one's comatose. The same lady with the ruptured embankments, the rivers creating havoc in her brain. She'd been given molecules of Phenytoin, (which Con himself had handpicked from the trees near the forest in his backyard) to silence the electrical sparks, the hyper-synchronous discharges in her brain. Strangely for some reason the Phenytoin molecules created havoc in her liver bed. Islands of Hepatocytes like tender rose saplings, trampled and uprooted, destroyed. Her bile channels are blocked and bile backflows into the systemic circulation reaching her eyes, which stare at you like the setting sun. Fulminant hepatic failure due to Phenytoin they called it.

Hepatology on call-

Any liver problems guys! Trudging daily through the hospital,

Searching…searching yellow eyes…

Cardiothoracic surgery…intra op scares and post op yellow stares. Labor room…yellow third trimester eyes and transverse lies. Psychiatry…eyes turning yellow in someone already blue? Emergency surgery…aha! We'll show you what we can do!

Biliary fistula due to a bullet injury? We'll fix it proper…don't you worry! A tube in his torn bile duct and nature handles the rest. Hmm…speaking of rest …I hope you aren't too tired?

I wonder why the evening sun's so jaundiced. Is night a must like death? Wait a minute…don't push that tube into its malignant duct! Cholangitis…sepsis…you'll give him all the works.

Let him rest in peace…jaundice or none, lets not create a faster exit for the sun.

Hepatology Ward-I have all to myself a garden of tender roses. Each day I water them and discuss management strategies for a faster recovery…all the latest works with two of my junior residents. They do most of the work off course! Each day…collecting dung, trimming leaves and so on. All I need to do is encourage them and encourage my withering roses so that one-day, at discharge they blossom and bloom. Eating well, sleeping sound and no more of the messy blood vomiting please! I have seen those varices safely banded and bile ducts surely stented. Bye for now…till another bleed, another fever or jaundice.

Chapter 86

THE TRAIN JOURNEY 1-3

JOURNEY 1

Abraham is a resident in surgery and is one of June's innumerable silent admirers. He is nowhere as handsome as Kris or Seshu but he truly loves June knowing very well that June doesn't care much for him. He once confided in one of his friends about his love for her upon which the friend gave him an interesting advice on how to monitor her feelings for him. "Keep looking at her pupils. If they are dilated you know you have made it."

Abraham could never once notice June's pupils getting dilated in his company. Once he tried taking a good look at her pupils on the canteen dinner table while filling up her glass of water and found them constricted.

After June flunked her exams Abraham would join her on her long evening walks sometimes appearing suddenly out of the blue near a crowded city intersection. One day he visited her in her room and after talking about umpteen non specific things they started making love in a fit of frenzy. Abraham forgot to notice her pupils this time. After her exams, which June sailed through in her second attempt she became a fellow in Hepatology as a stop gap. She was also considering various subsequent career options when one day Abraham barged in with a newspaper clipping that declared vacancies for a physician and surgeon to Antarctica. Before June knew it they were on the train to Goa. The National centre for Antarctic and Oceanographic research in Goa was the venue for the interview

June's Diary

4th September, 2002, 3:00PM. On train... glad to have a reserved seat to Bombay. My friend Abraham is a heavy smoker and my dread of getting caught in the habit, especially to kill time keeps me writing, scribbling making notes of whatever comes across, sometimes even what my friend relates.

Abraham is a surgeon and he has pretty interesting anecdotes to share, the slip shod histories they take in the local nursing home, general one liners hurriedly jotted down in skimpy files, sometimes so very inadequate that one has to ask the patient before operating on him as to what operation he was planned for. Abraham used to be a good reader in his undergraduate days and now he just can’t stand the site of a book, any book, preferring to blow smoke out of the train window gazing vacantly at the countryside rushing by…does a surgery residency do that to you I wonder?

JOURNEY 2

June's Diary

3:15PM…the train leaves Kharagpur. Sounds of trinkets and clapping…a mental picture of Eunuchs, a vague feeling, almost a wish they may be just girls after all. As they approach us one of them keeps a hand on Abrahams head. I dish out 10/- feeling threatened with the look of fear in Abraham's eyes.

6:00PM…train window. The rain swept landscape, a black sky, green washed paddy fields and red huts with red mud roads to match.

A few rock outcrops sprouting from the green…as if peacefully grazing animals, some of them with shapes of domesticated dinosaurs, ponds which take on the color of the surrounding green and a host of kaash flowers booming amidst all these.

A name to this place, station Sini flashes by…all trains doesn't stop here. Like the poet Kedarnath Singh I don't wonder why then does this station exist at all. A few wet lands…marsh lands…a black elegant Santhal woman walking on the red mud road towards the village…a game of football in a large green field in progress, towns and trees suddenly sprouting in the green…a blue hill in the background, joining with a few larger ones further down and as the mountains approach nearer and nearer they start dominating the train window.

The train nears a larger town, Chakradharpur. Its power station houses a few Electric poles that somehow resemble alien robots (with all its elaborate ceramic work) watching us while we wonder when they'll come alive. On the other window a red temple top peers out from the green. The sky has reached a state where it's breaking up into myriad colors of blue green and red.

8:00 PM, Buying water bottles lavishly each and every time we feel thirsty instead of filling it up like old times when we were undergraduate students and traveling on the general compartment. It's not always that we couldn't afford a reserved seat at that time but more often our journeys would be planned all at the last moment.

5th September, Woke up with thoughts on the ancient Harappan civilization 3000BC? River Saraswati dried up 2000 BC? The spinal fluid in the central canal too dries up at the age of 40. City hutments, factory chimneys, a trail of sigmoid smoke across the sky... Telegraph poles stand in a single file.

8:00 AM Nagpur station: bought a newspaper…lots of beggars with cut hands and deformed fingers (? lepromatous). Ignored them like a thick skinned rhino while they carried on their incessant whining…initially extolling your virtues but as they despair at your indifference, finally leave showering the choicest abuses (which are thankfully garbled and difficult to decipher). A lot of children sweep the train compartment free lance and hold out their hands seeking compensation for their bit of work. It's probably better than begging (which angers our traditional values) but then it encourages child labor. An old man free-lance sweeper appears. He looks like having shifted from the begging school to the more techno ideal savvy money for work school. However his demeanor is more of a beggar as he whines for his money from everybody in the compartment holding out his broom as a proof of having finished his job. There was hardly anything left for him to sweep by the earlier children who preceded him by a few minutes. You wonder if he follows them on purpose.

JOURNEY 3

June's Diary

9:00 AM: Talked to a young fellow passenger on the window opposite mine. Stays and earns his living in Bombay. Started working in silver chains after studying till the 6th standard at his home-town Arambagh. Later after 3 years he was taken by a relative to Bombay and has been working on gold chains ever since. His day starts at 6:00 AM and ends at 12:00 AM at midnight. Really! To think doctors had all the work. Initially he was an apprentice for 2 years and only recently since the last 6 months has been confirmed with a salary of 5000 per month…much more than what his elder brother gets, 500/-per month for fixing gas cylinders.

9:20 AM: The beggars of Bombay have a more violent attitude according to Abraham. They aren't just content with abuses for non-givers but even resort to violence if particularly ignored…more of dacoits than beggars. Do they have a mafia backing, we wonder. As I write … a crutch glides along the recently swept compartment floor followed by a man with an amputated thigh and eyes begging notes.

10:05AM Rows of crops looking like a huge kitchen garden. Abraham feels its cotton. The sun smiles pleasantly at them…not a single cloud in the sky. At 12:00 noon it's going to be real raving mad HOT. The train stops at a station and a beggar peers at our window. He's thrilled as Abraham absent mindedly hands him a 5 rupee coin. Looking ecstatic, he keeps on showering praises on Abraham…a generous beggar in terms of praise.

10:20AM a blind flute player with an out of tune flute. Again begging with a flavor added. The woman with two small children lying lazily in the middle berth finally climbs down throwing nervous glances as she dangles her legs.

10:30 AM: Tomato soup…Abraham's incessant capacity to keep guzzling. The price of most food items is 10/-, be it a packet of wafers, tomato soup or a bottle of water. Two stations with peculiar names Barabamboo and Retard flash by.

4:30 PM: Abraham's story, an anesthetist beats up somebody and throws him off the bus and later after he reaches home gets a call for an OT case. Goes and finds it was the man he'd clobbered.

5:00 PM passed a few mountains looking like Mesas…nearing Nasik, a good distance from Manmad.

6:00PM Bombay VT, A long queue for the train to Goa...no choice except the general compartment. Surprisingly clean for a general compartment. Joined people sitting on the floor…wished we had kept the newspapers. After sometime I found myself rolling off on to a peaceful sleep below a berth. Sleep knows no barriers...once asleep there's no difference between the soft cushions of first class and the general compartment floor. This philosophy is ill spent on Abraham who considers it below his dignity to sit, let alone sleep on the floor.

He's understandably mad at me when I greet him early in the morning commenting how much he resembles a sleeping horse. The train has meanwhile stopped near a tunnel and there's a perfect dawn breaking out of the window. The Western Ghat looked a welcome green behind a mist peeling off. The dark clouds atop the hills promise rain.

Chapter 87

ON BEING A MEDICAL STUDENT 1

Abraham proposed to June soon after they received a sheaf of papers from Goa informing them of their having been selected for Antarctica. Their jobs off course would have simply involved looking after minor ailments in the crew. As she was expecting the question, June had already researched her answer for the interviewer when he asked her why she wanted to go to Antarctica. "I understand life in Antarctica isn't going to be comfortable and I want to see how much discomfort and adversity my system is capable of withstanding." They were less impressed with Abraham who said he wanted to touch the penguins. One of the interviewers glumly remarked he had seen a penguin die after having been touched by a tourist. He also asked him wryly if Abraham would go even if June wasn't selected to which Abraham fell silent and finally uttered a painfully strained yes.

June's answer to Abraham's marriage proposal was in the negative. Abraham had added that they would be the early couples in the world to get married in Antarctica. Abraham had somehow taken for granted that June was his possibly because they had slept together. However she was far from the usual stereotype of the media hyped Indian woman. June at first expressed her resentment against the fact that this was a Government job and they would have to go through a load of red tape and bureaucratic hassles. The first inkling of the hassles was this initial installment of a sheaf of papers in which they were required to fill in all their details for the umpteenth time. "Abby baby I admire and like you but I can't love you enough to marry you."

June shoved an opened journal page onto Abraham's glum hands. "Take a look at this article. I have already received an offer to work with the man who's written it.

Abraham glanced at the journal article June had shoved into his hands. It was written by some VM Hegde, Prof of medicine and Dean, Macchapucchare college of medical sciences, Pokhara, Nepal. It was more of a narrative essay rather than a so called scientific article.

I had started being a medical student 40 years back and 'am part of the teaching faculty now but I still feel I am a medical student. The science is ever changing and whatever we learn today soon is washed away in a deluge of rapidly accumulating newer evidence. We tend to look back and think how awed we were by our seniors who always seemed to know so much. We thought that we too would know as much when we got to be senior ourselves. Many years rolled by and that moment never came, only we noticed ourselves being thought of as more knowledgeable by our juniors. Whatever we learnt instilled in us, the capability to apply a rule of thumb in patient management. However, it never worked when confronted with diagnostic uncertainty complicated all the more by a dense jungle of evidence, which has grown remarkably over the years.

Chapter 88

On Being a Medical Student 2

As medical students, when we finished the basic sciences and started clinics, there was a pleasurable difference noted immediately, as we no longer had to cram dull theory like in the basic sciences. The patient was our greatest teacher of medicine as much as the dead body had been in anatomy.

The best way to learn was interviewing the patient and going to the depths of his story, which in 75% of cases would yield the diagnosis. It was a detective game and the clues had to be meticulously elucidated. This is where our teachers played a role. They showed us how to elicit these clues, which were vital to the diagnosis. We however learnt not only plain hunting but also to love our patients. It was great to collect their stories and keep them in our mental hard discs and reflect on them over our textbooks. That was the first time we experienced our books coming alive. At a time when our contemporaries were collecting stamps, coins or people's autographs, we started collecting people… live people who were not just long dead characters of a novel but would greet us from bus stops or morning walks. Interacting with them we experienced a vitality flowing like a river in and out of our lives.

The greatest gift however, my formal training in medicine provided, was the ability to look at the human body as a machine. I had initially instinctively rejected this mechanistic way of looking at humans but gradually became used to it when I realized it was one of the most advanced systems ever invented or evolved. In this scheme of things, off course, doctors lose their halo and become more of system troubleshooters. They can be mollified by the fact that its one of the most advanced and complex systems they trouble shoot.

However it's peculiar how a whole health care industry manages to remain preoccupied with only one single objective day in and day out and that is system trouble shooting. We don't develop new systems, we simply try to solve problems and keep learning from them, newer methods, and strategies to solve further problems. Very often there are problems in the system we have created ourselves in our bid to solve a few relatively simpler ones. Luckily the body in its advanced complex framework has been endowed with self-repairing capabilities. Its system of self-repair often also does lead to problems, a reminder of the fact that we live in an imperfect world.

All humans have this propensity to solve problems be it mechanical or human. There is a medical student in all of us. More often we tackle our innumerable day-to-day problems just by listening to them, analyzing them from a different perspective. This however requires time, something, which has been lost to medical students who are also, designated interns, residents or consultants.

June had emailed Prof VM Hegde thanking him for writing the article and the professor in turn had enquired politely on her present engagements. On learning that she had recently finished her residency from the IIT (Indian Institute of Trouble shooting humans) he immediately offered her a job in his own Institute. You'll love the place. The mountain after which our college is named dominates the skyline. It's even grander than Matterhorn from Zermatt.

Chapter 89

MACCHAPUCCHARE COLLEGE OF MEDICINE

June's Diary: January 30th

It has been a hectic but pleasurable first few months working as a Lecturer, Medicine in this dream of a medical college in Nepal.

The first room in the medical ward as soon as one stepped out of the ICU was also called the Macchapucchare room as one could see stupendous views of the snow covered conical mountain towering over the hospital, barely 35 kms away as the crow flies. My eyes sometimes tended to rest on it even before I glanced at the patient.

This was an 80-year-old young lady who had been having high grade spiking fevers and initially had been diagnosed to have a urinary tract infection as the urine culture had grown E coli. She had been put on a program of Gentamycin as the bug was resistant to other small fries. However she went on spiking merrily each time touching base line, the temperature sometimes falling below it. An ultrasound abdomen revealed a liver studded with multiple abscesses along with gall-bladder calculi, a dilated common bile duct and its ramifications into the liver.

Some of the gall bladder stones had fallen into the cbd pipe and were obstructing her bile flow. A few otherwise innocuous bacteria had seized the opportunity and got into the bile pipes. Then they had clambered up the pipes into the liver cells. Within a few days they had freaked out on an orgy of destruction and self indulgence, clearing away dense jungles inside, building homes, having families and children. The 80-year-old would have all this while been blissfully unaware of the goings on and one fine day when her blood defense systems woke up to an alarmingly expanding bacterial colony she experienced the first spike of fever, her appetite replaced by this new uncomfortable feeling of nausea. We urgently needed plumbers to remove the gall stones obstructing the drain pipe. They would also keep an artificial tube there possibly right up to the abscess and try to drain the masqueraders out.

June's Diary February 4, 2003

This is a weekend and I usually get to pen down my thoughts on patients I see over the week only during this time. I had just woken from a pleasant dream on Monday morning. We had crossed the hanging bridge over the river Seti flowing like a torrent in front of the faculty quarters and trekked up a wooded university campus to the Bagar bus-stop. On the way we passed a few trees labeled with badges showing their scientific names. They looked like they were distinguished personalities with badge-labels showing their name and designation. A long line of black ants were creeping up on their bark and we peered intently at them trying to figure out where they were headed. They seemed to be vanishing into the intricate branching arising from the stem. Do our thought processes too vanish into an intricate network of our decision trees?

Chapter 90

SETTLING DOWN TO A NEW DREAM

June's Diary Feb 4th 2003

The bus isn't crowded like a regular city bus. A few passengers get down at a nearby stop. A new passenger gets up and eyes me with a hint of recognition. One of my patient's relatives I presume.

The next bed in the special ward is a 90-year-young Second World War veteran. It has been years since he left the army and came down to his village, enjoying the gentle comforts of rural life. His abdomen was distended with fluid in the sac covering his intestines. This sac was made of a thin layer of cells and had been also covering his lungs since he was a fetus. At his age of 90 as the fluid in the sac covering his intestines increased, water also seeped into his lung coverings which were in continuity with the intestinal coverings.

The cause of all this water accumulation was off course in his liver that over the years of dealing with a generous supply of Alcohol had fibrosed…or did it just age like all other cells of his body? No...Aging wouldn't cause fibrosis. Whatever it was, all this fibrosis was compressing the pipes supplying blood to his liver. The rivers inside those pipes were trying to find other avenues of drainage by trying to percolate into his visceral coverings. We thought it was alcoholic cirrhosis.

We questioned him on his alcohol habits. He said he did take alcohol and with a wink added it was just to taste her daughter's home made local brew to make sure the taste was all right for their guests.

As the bus drove into the lakeside we made our way through a variety of stores selling items of tourist interest, trekking shoes, sleeping bags, instigating us to take off into the wild, away from this busy market place.

The lake was calm with a few boats gently drifting in its midst. There was an island which too seemed to be drifting. A few rain clouds had descended on the adjoining mountains full of lush green forests, a pagoda stared at us from its top. We watched our surroundings from a comfortable vantage point near the lake, inside an open air restaurant, a few buffaloes having a dip, children diving in to join them with a splash.

June felt relaxed swimming with her colleagues after having taken the plunge into her new job. She was a consultant trouble shooter now and a teacher-lecturer at that. Their hospital didn't lack in system problems. Solving these were at the same time less taxing as the number of patients wasn't very high like in city hospitals. She put all her heart and soul into dealing with them as best as she could and reveled in the questions asked by her students, an important way to keep learning.

Chapter 91

THE STORY OF GLOMERULAR INJURY 1-7

GLOMERULAR INJURY 1

It has been long since June met Professor Joatmon in his limestone cave of holistic medicine full of stalactites and stalagmites. At present if you move these curtains you can spot June looking smart in that suit, standing on the pulpit with a laser pointer in her hand. From a lecturer she has quickly become an Assistant professor now, teaching in the Macchapucchare College of medical sciences, Nepal. She's at the moment presenting a paper on preventing and managing glomerular diseases in the community. You can stay on if you'd like to view her presentation.

June starts off with, "May we have the first slide please?"

Slide 1. Macchapucchare teaching hospital, Pokhara, Nepal.

The first slide is a picture of Macchapucchare teaching hospital, Pokhara, Nepal where I live and work. To talk on renal diseases their impact and strategy for long-term solutions I am going to begin by relating a story, which has been synthesized from the stories of a number of our patients and doesn't necessarily reflect any real life patient. Prem Bahadur Khadka (not his real name) is a 25 year old boy who hails from a remote village near Jumla, Nepal and had never seen a bus or car in his life before he left his village for higher studies.

Slide 2.

This is his village where the only means of travel was an airstrip apart from the other option of a few days of walking.

Slide 3. Professor Joatmon's village.

Prem was intelligent and soon after he finished his 10th grade exams he decided to leave his village for higher studies.

GLOMERULAR INJURY 2

Slide 4.

He came to Pokhara, enrolled in a good school and saw proper roads and vehicles for the first time.

Slide 5.

As he was a good student he got into engineering and after finishing his degree, arrived in US as a software analyst. However he felt miserable there as he kept missing his relations and his mother's cooking (among other things). He was actually relieved when he got the pink slip and lost his job after the economic slowdown.

Slide 6.

He came back to his village (and his mothers cooking) but found that by now he couldn't adjust with his brothers who knew more than him about farming and all his knowledge of software was of no use in the village (which didn't even have a single computer).

GLOMERULAR INJURY 3

Slide 7.

While all this was happening to Prem our protagonist in his macrocosm, quite unknown to him or anybody else a major war had already started inside his kidney where his glomeruli were being attacked by a lot of inflammatory cells.

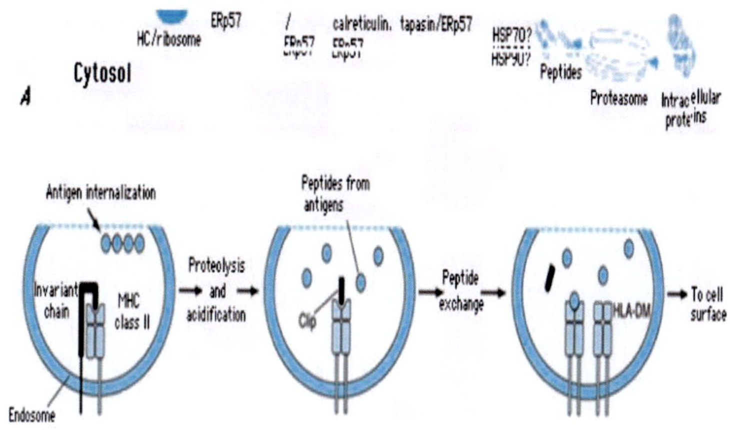

Slide 8. A cell in Prem Bahadur's glomerulus.

This is a cartoon borrowed from Harrison's principles of Internal Medicine, which shows in molecular detail the happenings inside a cell in Prem Bahadur's glomerulus. You can see how the antigens are processed and finally parceled into an endosomal compartment containing the MHC class II molecule .The antigen settles down into the groove of the MHC and sets off a chain reaction stimulating a clonal proliferation of hordes of inflammatory T cells.

GLOMERULAR INJURY 4

Slide 9.

This is the strong inflammatory response inside Prem Bahadur's glomerulus as a result of which we expect to see a lot of protein and RBCs in his urine (if only we could have examined it earlier). However Prem Bahadur didn't notice anything wrong with his urine. His neighbors noticed him growing plump day by day and complimented him on this sign of prosperity. At first Prem Bahadur was also happy that he was getting fat but later noticed that he was unable to enjoy his mother's cooking due to a feeling of extreme nausea. This too was ignored by him until one day he developed extreme breathlessness and had to be rushed to our hospital in Pokhara.

Slide 10. Our hospital in Pokhara.

This is a picture of our hospital on a stormy night. When I saw him in our casualty (for the first time) I found him gasping for breath, his lungs were full of crepitations which we quickly treated with Lasix and referred him to Katmandu for dialysis. At Katmandu they dialyzed him for fluid overload and a urine analysis subsequently showed 3+ proteins and plenty of RBCs suggesting acute glomerular injury. He was immediately treated with high dose steroids and a renal biopsy was done.

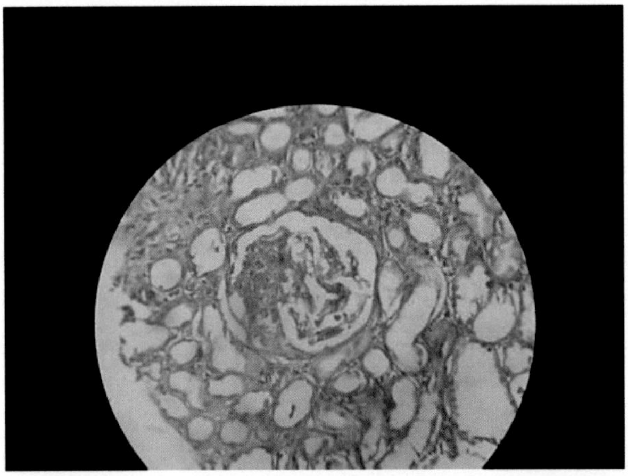

Slide 11. Renal biopsy.

The biopsy showed focal areas of glomeruli that were sclerosed in segments.

GLOMERULAR INJURY 5

The fire was too severe to be quenched with steroids and it was already too late to prevent Prem Bahadur's kidney from blowing up in smoke.

Slide 12.

He received a few more dialysis from Kathmandu and spent whatever money he had earned over the past few months. His brothers came forward to donate their kidneys for transplantation but that would mean selling off their land and cattle to go to one of the hospitals in India and Prem Bahadur wasn't keen on that. The next slide is a Haiku sequence, which was mailed to me by a friend right after 9/11.

breaking news
into the darkness
a shattered tea cup

flags aflutter
the still smouldering debris
-just another day

different skyline
through the settling smoke
some more sky

each passing hour
through broken windows
only silence

Slide 13. Haiku sequence.

GLOMERULAR INJURY 6

The last I heard of him was through one of our medical students also from the same village. Prem was spending his last days in deathbed with his family praying for him daily and it would be a matter of days or months before death would take him.

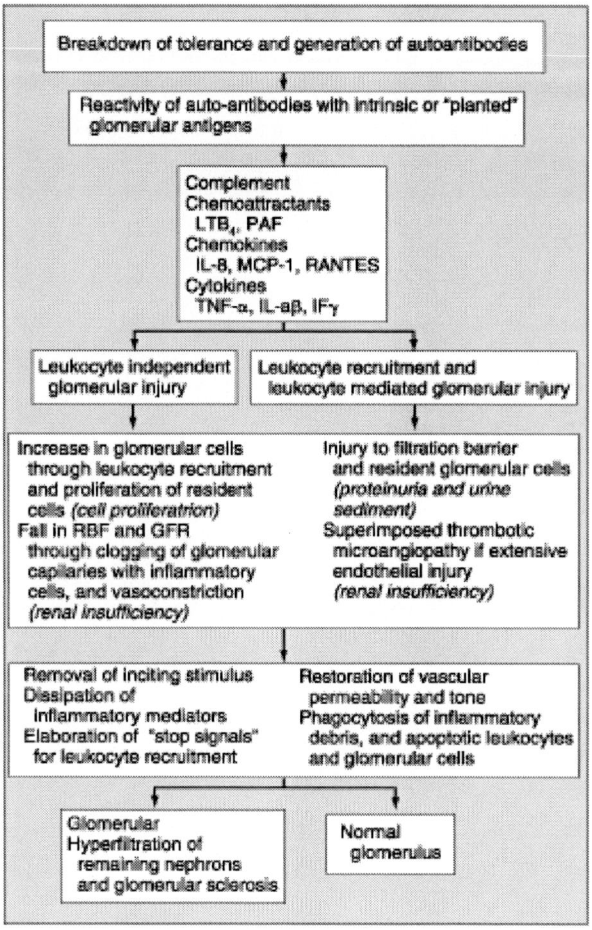

Slide 14. Summary of the story of glomerular injury.

This is a summary of the story of glomerular injury taken from Harrison's principles of Internal Medicine which describes how it's initiated by breakdown of tolerance leading to reactivity of antibodies with planted glomerular antigens which in turn generates a chemical cytokine mediated response leading to proliferation of inflammatory cells and subsequent irreversible renal damage (in a substantial number of people).

GLOMERULAR INJURY 7

The key word here is tolerance and (next slide).

Slide 15.

Tolerance is fast becoming popular in transplantation research but we need tolerance urgently before the need for transplantation arises to save our native kidneys (before the inflammatory cells attack them). As much as we need tolerance to save this Earth before a full-fledged war breaks out. That is the problem of glomerular injury for we don't really know why this breakdown of tolerance occurs. We don't even know why hypertension or diabetes occur (for that matter). Our present strategy is only to control them in the hope of slowing the progression of renal disease.

Chapter 92

GOING UP AND DOWN TOPS AND BOTTOMS

June had begun the most crucial learning phase of her life that Con was looking for. There were two traditional approaches to learning, the top down and the bottoms up.

Like her friend Paul from the complexity list elaborated in an email,

> "Top down is a pre-formed idea and we can choose to say yes or no. Bottom up is an exploratory idea where new direction can be woven from multiple diverse strands (many of which can be "top down"). Top down is essential; bottom up is existential. Top down relates to realist ontology - it is really objectively there; bottom up relates to relativist ontology - it is there if it is meaningful to you. Today's bottom up can be tomorrow's top down and vice versa. Top down in one place is bottom up in another. The two are constantly (and rightly) interwoven at all levels.

June's undergrad years had been more top down assimilating her tools, a knowledge base accumulated over decades of bottoms up research by physicians of yore. Now was the time to use the tools she had acquired to fix human systems independently in a judicious mixture of the two approaches".

Jung Bahadur Gurung was a retired Indian army soldier who came into her outpatient chamber with a complaint of fever for a month. For a number of years he had had severe pain abdomen coming on intermittently for a number of years. He had yellow eyes that looked quite unwell and he kept scratching his body. June did a quick system river check and his blood tests indicated an inflamed liver evidenced by the enzymes. It also suggested the inflammation was obstructing the outflow of bile from his liver and this was producing all the scratching. June's gut feeling told her it was a granulomatous inflammation of the liver possibly due to tuberculosis.

Jung Bahadur was admitted for a liver biopsy after June asked her radiologist friend to look for and rule out any dilated bile pipes inside the liver that might suggest a block to the pipes carrying bile into the liver. He took a quick look with an ultrasound beam and reassured her that the liver pipes were all right. June went ahead with the liver biopsy and got her liver tissue but was shocked to find that she was hitting pus. That night the pressures inside Jung Bahadur's system rivers hit an all time low at 60 mm of mercury even as his temperatures swung up to a high of 40 degrees centigrade and June thought she would lose him.

The commonest complication of a liver biopsy is bleeding from the liver bed but June was puzzled, as his hemoglobin was not dropping. A repeat ultrasound examination revealed a large sub diaphragmatic abscess that had been overlooked in the previous ultrasound. June realized she was partly responsible for this as she had hurried the radiologist into simply ruling out a specific top down problem and not having allowed him a complete bottom up look. Jung Bahadur was operated for the abscess below the diaphragm and the surgeons also discovered a perforated gall bladder. The pus had obviously come from an infected gall bladder and the organism was found to be Salmonella paratyphi. Salmonella getting into the blood stream was a common cause of fever in this part of the world but this one had got into the gall bladder, perforated it and caused an abscess manifesting as prolonged fever. Jung Bahadur followed up with June for years to come after his complete recovery.

Chapter 93

Fever Bugs in System-Rivers

Apart from the unusual presentations of fever the usual fevers were no less interesting. Typhoid was endemic in the area, possibly as compensation to malaria that was less common. Typhoid meant, 'like Typhus', which was less common but present nevertheless. Systems afflicted had this splitting headache, apathy and loss of appetite. Most systems malfunctions could be classified into that affecting the system from outside (bugs getting in through online attachments) or from inside due to the systems anti-bug defense software gone awry. June got to see more of system problems due to outside bugs.

Fever was indicative of the anti-bug software working overtime, producing chemicals like lymphokines that when dumped into system rivers would naturally heat up the system. The commonest bug to cause seasonal fevers was off course the common cold virus where systems would start blowing their nose along with sneezing and dripping secretions everywhere. Most of these malfunctions would be temporary and would subside on their own hardly requiring any anti-bug software. The Typhoid bug on the other hand would also produce high grade fever that would characteristically refuse to touch normal baseline in the first few days. June found monitoring system temperatures by plotting a 4 hourly graph quite useful in differentiating between viral and typhoid fevers at least in the initial two or three days. It was all the more necessary to prevent misuse of anti-bug software in the community. In fact anti-bug misuse was assuming epidemic proportions in the community far greater than the bug epidemic.

Once anti-bug software was used frequently the bugs would get to recognize their weak spots and modify their codes accordingly so they could escape their next onslaught. Most community practitioners, health assistants and physicians alike resorted to prescribing antibiotics under the mistaken assumption that they would bring prompt relief. June found it was paracetamol that would be able to give them better relief. Anti-bugs more often would just add to the misery by producing diarrhea. She began to design a prospective trial in the community to see if distributing fever charts to patients could really help to bring down this present epidemic of anti-bug misuse.

Apart from the common typhoid, June noticed another particularly different pattern of fevers, as high grade as typhoid. It was however intermittent in pattern even in the very first days of its appearance and was remarkably resistant to anti-typhoid software. It would come in epidemics every monsoon and most people affected were farmers. They had biochemical abnormalities in the form of raised enzymes in their liver function tests and mild degree of proteinuria and WBC casts suggesting a liver and renal involvement. This was Typhus which responded remarkably to another software, doxycycline. The name Typhoid was actually derived from Typhus meaning typhoid like and yet Typhoid ruled the roost as an endemic disease while Typhus was just an epidemic visitor.

There were systems met infrequently that would present with continuous fever where no amount of investigations would reveal any clue. Very often they would demonstrate multi-component involvement, be it in the joints, or skin, or kidney filters, or heart, lung coverings or even the bone marrow affecting their cell producing assembly lines. Here it was the systems own anti-bug defense gone awry attacking and corrupting own files. Most of them responded to software labeled immunosuppressive. Sometimes they would respond to nothing.

Chapter 94

THE CRANKY IRRITABLE BOWEL

June's Diary (Draft of a Mail to Jo- on Her Experiences with the Irritable Bowel)

My first acquaintance with the disease was possibly at a time when I didn't have any connection with medicine. It began with my father. He was a very meticulous man and was by profession an anthropologist-teacher. His abdomen at that point of time while I was in school oblivious of my future medical career was gradually putting on a paunch. It was not however his paunch that bothered him, it was gas. Gas always used to have ticklish connotations for us young ones probably as it came out from banal orifices producing an offensive sound and odor. However it was not the coming out of gas that seemed to bother him. It was the not coming out part that was most worrisome.

He used to feel miserable about his intermittent abdominal bloating, producing a diffuse pain. Very often he would express his anguish by percussing various parts of his abdomen, even while moving around with his daily activities until it became a habit. He even developed a philosophy of happiness which stated that all cause of human misery lay in the abdomen. If only abdominal organs functioned well, everything would be perfectly alright in this world. Surprisingly he grew out of it although it took him years and I no longer find him percussing his own abdomen.

I went on to meet similar people in my human trouble shooting career. At first I used to attribute them to protozoan infestations like Giardia or Entamoeba so very endemic in our part of the world but gradually realized that most often no cause could be detected. There were so many criteria to identify the irritable bowel and most of these problem patients would satisfy quite a few. The treatment was however dismal as the cause wasn't clear. It was a disturbance of bowel movement alright but then why should the bowel move so mischievously at times? A typical scenario was the bowel moving in alternate contraction and relaxations and suddenly at one point of time stopping randomly followed by distension of that halted segment until the human carrying the bowel winced in pain.

Mercifully it would pass off after some time, the distended colonic gas coming out silently or otherwise and you'd find a relieved expression on his face. Another way it would present was an urge to defecate at times especially when the subject was engaged in some kind of work, tedious and boring but important nevertheless. However what would come out could hardly be called fecal matter rather than that which was normally present on the surface of the colonic mucus membrane, 'mucus.' Each time one passed stools one felt there was more to pass, a sensation termed tenesmus that was possibly due to colonic spasm. Eighty of hundred patients in my outpatient clinic with abdominal pain going on for a very long time could be classified into this pattern.

Chapter 95

A DYSFUNCTIONAL DYSPEPTIC STOMACH

Vatta- an Ayurvedic concept: People with predominance of disorders of movement or motion in the body

Sutra

It was not just the lower bowel whose movement was at fault but even the upper bowel along with the stomach would refuse to budge at times. This produced an upper abdominal epigastric discomfort frequently labeled dyspepsia. June would put in flexible telescopes with lighted lens into these mischievous bowels either from above or below to look for anything that could be treated but no it was something to do with the electrical wiring of their slow moving pipes that wouldn't move because of frequent current failures. There was no way she could have their wiring replaced. What could have led to their movements getting awry? She decided to drop a mail to her old professor Joatmon.

Dear June, Sitting in a busy OPD (with both patients and students) and evaluating dyspepsia (which happens to be one of the commonest patient complaints in the OPD closely rivaling headaches and chest pain) one needs to develop a quick but full proof strategy to evaluate these patients. We had been taught a lot of clinical features to distinguish between duodenal and gastric ulcers in our undergraduate days but what we possibly needed more was to distinguish between ulcer and non- ulcer dyspepsia.

Dyspepsia Subgroups

Reflux-like dyspepsia
Heartburn plus dyspepsia
Acid regurgitation plus dyspepsia

Ulcer-like dyspepsia
Localised epigastric pain
Pain when hungry
Pain relieved by food
Pain relieved by antacids or acid reducing drugs
Pain that wakens the patient from sleep
Pain with remission and relapses

Dysmotility-like dyspepsia

Upper abdominal discomfort (pain is not a dominant symptom)
Early satiety
Postprandial fullness
Nausea
Retching or vomiting
Bloating in the upper abdomen (without visible distension)
Upper abdominal discomfort often aggravated by food

Chapter 96

SETTING GAS PURELY UNCONDITIONALLY FREE

There are a lot of clinical subgroups initially devised in the hope of achieving therapy tailored to specific symptoms but they don't help in distinguishing clinically between ulcer and non ulcer dyspepsia. In fact the only thing, which might give us a clue to an ulcer or an endoscopically verifiable structural lesion, is the presence of recent onset dyspepsia. Patients having symptoms of long-term dyspepsia do generally turn out to have no ulcer disease on Endoscopy. In a Western series, only four out of 165 dyspeptic patients who did not have a peptic ulcer at initial Endoscopy developed an ulcer over 10 years of follow up, although 62 (58%) of the 106 who still had dyspepsia 10 years later had ulcer- like symptoms. What then is the cause of this dyspepsia, which is so very distressing to our patients?

If most of them are supposed to have ulcers only when the dyspepsia is recent could it be that the ulcers heal and leave no trace of organic lesion but do cause some neural damage on the long run affecting gastric/bowel function. In many patients, failure of the peristaltic wave to propagate results in upper abdominal discomfort, early satiety or post-prandial fullness. There are no medicines yet in the market to really improve bowel motility (Cisapride, metoclopramide and domperidone aren't practically that effective) , however there may be some benefit in walking after meals, at least this has been demonstrated to improve gastric motility in diabetics. It's rather surprising that it hasn't yet been tried or verified in non ulcer dyspepsia.

There are many questions, which remain unanswered even in ulcer dyspepsia. Like why does H. Pylori eradication only improve ulcer healing and doesn't have much benefit in non-ulcer dyspeptics who harbor H. Pylori? Is it because H. Pylori is not really the causative agent of ulcers but only responsible for delayed ulcer healing?

I would suggest that you counsel your patients to optimize their bowel movements such that their bowel doesn't get confused. All they need to do is move it everyday at particular fixed times. In other words eat and pass stools at the same time every day. As these bowels are autonomic and do not directly take orders from the brain they can only be regulated by timely stimuli or signals to their electrical apparatus. Walking regularly in addition would also offer a timely stimulus to their regular movement. The same could be inferred about sleep I guess but all these definitely need to be validated by controlled trials and why don't you take up the challenge?

Yours,

Jo.

Chapter 97

A Blackening Lady

One day, June noticed in her outpatient department a middle aged lady who appeared strikingly black, a complexion never noticed in her part of the world. She quickly asked her in and admitted her after a brief history.

She was a 40 year old lady from Mustang valley, Nepal and her main complaint was progressive blackening of her whole body since one year. She also had associated weakness along with loss of weight (>50%) and appetite. Since one month she had also developed progressive contractures of her knees and hips and was virtually bed ridden. She had taken some indigenous medicines for her troubles that continued however. The exact nature of the medicines could not be ascertained. She was also amenorrhoeic since a year. Fifteen years ago, she reported having fallen very ill once with cough and fever and was treated again by indigenous medicines. The illness had lasted for 1-2 months after which she recovered almost as if from death-bed.

On physical examination she was cachectic (weighing 30 kgs) and pale along with diffuse hyper-pigmentation of skin and dorsum of the tongue. The blood pressure was 84/60 mmHg in the right upper limb and pulse rate was normal. The abdomen revealed a mild diffuse tenderness and the knee and hip joints showed bilaterally symmetrical contractures. There were no other remarkable findings on systemic examination. Her Hemoglobin was 11g/dl and total leukocyte count was 6,200 with a normal differential count. The serum sodium was the lower limit of normal at134 mmol/L and the serum potassium was 3.3mmol/L. The liver function tests showed a mild rise of enzymes. The chest X-ray showed a small calcified focus in the left upper lobe of lung. An ultrasound revealed a right suprarenal hypo echoic mass posterior to the Inferior vena-cava, measuring 2.5 X 1.8 X 1.6 cm.

Displaced retroperitoneal fat planes seen between the mass and adjacent liver parenchyma suggested it was an adrenal mass. There was no similar obvious mass in the left adrenal area. An ultrasound guided fine needle aspiration was performed which showed frank pus. This on further microscopic examination showed gram negative bacilli that on culture grew mucoid colonies indicating Klebsiella pneumonia sensitive to ciprofloxacin, Gentamicin, Cephazolin, ceftazidime and resistant to Ampicillin.

The patient was already put on Anti- tubercular therapy after a clinical diagnosis of Addisons disease along with steroids. After the report of the pus culture sensitivity oral Ciprofloxacin was also added for 2 weeks. On follow up after 6 months, the patient reported an increase in appetite and strength, the blood pressure was 100/70mmHg on the average along with a 4 kg weight gain and the contractures improved with physiotherapy as the patient gradually became ambulant. A repeat liver function test showed normalization of the enzymes level.

Chapter 98

A WEB BASED DOCTOR PATIENT RELATIONSHIP

A WEB-BASED RELATIONSHIP 1

The patient gave June and her colleagues a lot of surprises. Although C.T scans are known to reveal enlarged adrenals at advanced stages of tubercular Addison's disease, it is not very common to visualize them on ultrasound, let alone do an ultrasound guided fine needle aspiration. To her further wonder it turned out to be pus, which revealed the bug Klebsiella pneumonia as a possible source, hitherto unheard of as producing Addison's disease. There were two possibilities, either it was secondary bacterial infection on top of primary involvement of the adrenals with tuberculosis or it was a primary infection of the Klebsiella Pneumonia involving the adrenal gland.

Odd for the bug Klebsiella was the remarkable lack of fever, or any other symptoms to suggest the portal of entry (unless it was the febrile pneumonia like illness she had had 15 years back). However a source could also have been from the urinary tract as Klebsiella is known to cause asymptomatic bacteruria. June posted her experience on a medical web site and one fine day was surprised to receive a mail from another patient long after she had put the whole thing at the back of her mind.

Dear June,

I read your case report "A blackening Lady". I found it interesting.
I am currently treating Klebsiella pneumonia bacterial infection.
Would you please respond with your thoughts on the following question...

I'd struggled with recurring symptoms related to klebsiella infection for the past 15 years. In the beginning of those years, I was suffering with recurring flu-like symptoms. Bronchial infections began then as well and continued to return (with the "current jelly sputum"). Urinary tract infections recurred often, a back problem began 7 years ago and progressively grew worse...I was told it was termed ankylosing spondylitis.

Here's my question...How long can an overgrowth of klebsiella pneumonia bacteria be present? Is it reasonable to assume that this has been the underlying problem for these past 15 years? According to the research I'm finding it would seem logical, yet I cannot locate any information regarding how long this overgrowth can remain in the body.

The one mention of that issue was stated "death occurs within 48 hours", I believe that is inaccurate information. Can you assist me with this question?

I'd greatly appreciate you're response. Chris Jestus, Minnesota

A Web Based Relationship 2

Thanks for your mail. Could I ask how you chanced upon my article? Before I attempt to answer your questions I need to clarify your symptoms in the history. Where was the klebsiella in your case isolated from? Was it the sputum or the urine? How many times has it been detected in your sputum/urine? Klebsiella can present in a variety of ways and our "blackening lady" was one of them. It can present as severe sepsis where the bacteria continue to divide relentlessly and can definitely kill a patient in days. However in your case its presentation seems to be what I understand from your history as a recurrent respiratory tract infection in which case it's unlikely that it will do much damage. Please let me know the details, particularly if there is any finding in your Chest X-ray. Thanks for writing and do reply for I want to help you solve your problem, yours sincerely

Juneli Sen. MD, Asstt Prof,
Dept of Internal Medicine,
Macchapucchare College of medicine, Pokhara,
Nepal. PIN -33701
Phone: 00977-61-26416-extn-189,
Fax-00977-61-27862

Dear Juneli, Thank you for responding.

I chanced upon your article by researching Klebsiella Pneumonia bacteria; I've been looking into many sights to confirm my belief that the Klebsiella has been the underlying issue in my situation. I believe, based on the research I'm finding, that it is possible.

The Klebsiella bacterium was detected through Comprehensive Stool Analysis, through Great Shakes Diagnostic Lab.

Regarding the sputum and urine... the sputum was never tested, yet in the past have had sputum that was very concentrated (almost a rubbery consistency, with a darker color)...recognizably different from the normal phlegm consistency. I've noticed this when the bronchial infections are most severe.

In past urine tests, E. coli showed up a couple of times. I'm seeing pictures that show a resemblance to klebsiella. A webpage that shows the similarities is http://www.mfi.ku.dk/ppaulev/chapter33/images/33-4.jpg I question if it's possible that klebsiella may have been present. Urinary tract infections had been recurring at least a few times a month throughout the years since then. I kept a good supply of cranberry capsules on hand, as they were often needed. I also consumed a lot of garlic throughout the years to fight off frequent illnesses.

My ongoing struggle with recurring symptoms began 25 years back. At that time my medical report noted " diarrhea, cramps, acute viral gastrointestinitis, Leg aches, fibromyalgia symptoms, blood tests showed neutrophils low (46.7) lymphocytes high (44.2), IgG antibodies". From then on ....I never felt completely healthy.

A Web Based Relationship 3

I had been prescribed many antibiotics from that time through 1994...at which time I started researching natural approaches to help control the many recurring yeast infections, thrush, allergies etc. (which I understand leads back to overuse of antibiotics).

The following notes are taken from my medical reports....

Over the past 15 years I was to doctors many times with complaints of overwhelming fatigue, depression, and bronchial infections that would last for weeks, subside, and then return within a few weeks again.

My blood pressure would usually be on the low side (98/36, 96/40, 84/60, 80/50),

I frequently felt ill with flu-like symptoms, often had low temps as well (97*, 96.4*, 95.7*, 94*);

I was having fibromyalgia symptoms (often aching in various parts of my body, which came and went, recurring in different areas with each onset, more noticed in the morning).

I had chronic constipation that began in 1993... For the past 10 years I could not pass fecal matter without an enema and when I did, I often suffered from hemorrhoids and bleeding from straining. I became too thin and was often told by my husband and friends that I needed to gain weight. I often felt weak and periodically got dizzy.

Over the past couple of years, I was becoming more forgetful and struggled with confusion and unclear thinking. I struggled with manic-depression as well for the past few years. I also struggled with Pre menstrual syndrome, a lot prior to my monthly cycle along with recurring headaches often.

I would get sore lumps under my armpit (assuming swollen lymph node). My breast had become continuously tender and painful for the past year or so.

I'd began seeing chiropractors in 1996 with lower back pain which progressively grew worse, and eventually affected my entire back and neck, was particularly painful on left side between shoulder blade and spine. I wasn't able to sleep well, my eyes were often red and I was told I had Ankylosing Spondylitis. There's a good study mentioned by the university of Florida relating Ankylosing Spondylitis to Klebsiella bacteria overgrowth http://www.chem.fsu.edu/learn/kleb.htm and http://www.chem.fsu.edu/learn/radio.htm

X-ray (1996) L5 sacralized as per chiropractor. In 2002 another chiropractors notes read "Chris has bone spurs @ the 4, 5, 6th lumbar vertebra, the 5th lumbar is fused with the sacrum". Range of motion has fluctuated throughout the years from 30% to 90%...stiffness came and went but continued to grow worse in general.

Bone scan (3-16-01) showed right increase activity in upper left costal vertebra junction at T3 level, noted possible degenerative activity, no significant abnormalities within the verterbal bodies along with mild scoliosis of thoracic spine.

A Web Based Relationship 4

Christus's medical history continued.

A CT scan (5-01-01) noted cyst within the left kidney.

Echocardiogram (5-4-01) no significant findings noted. Estimated ejection fraction is approximately 65-70%. I had been experiencing bouts of intense chest pain that made it hard to breathe, and would let up a little when I bent over. Recently I found mention of tachycardia at some sights for Klebsiella...tachycardia seems to describe my chest pain. This previous note of tachycardia symptoms was more severe than the often recurring general tightness/hard to breathe feeling that I experienced often as well.

In regards to helping me solve my problem...thank you for your interest...I sincerely appreciate that.

The above issues have all significantly improved. (I would say at least 80%). I now am sleeping well at night; the back pain is very minimal compared to before. The last bout of severe chest pain was over 3 months ago; I feel all the psychological issues have been gone since January, I am able to stay focused and my memory is better; I no longer get urinary tract infections like I used to, no headaches or P.M.S. symptoms in the past few months; I'm not aching in various places any more...I wake in the morning without pain. My eyes are not red all the time; I have no swollen lumps under my arm, my breasts are not hurting all the time etc. etc. In general...I've come a long way and I am so very thankful for that.

The improvements began to happen when I began passing long strands of mucous in my stools in November of 2002. I believed they were worms, but two different lab tests reported no worms. They were defined as "mucoid matter". (One test was done at a local lab, the other at Great Shakes Diagnostic Lab…the tests were done about 2 months after I began passing a lot of these strands) These mucous strands have ranged up to 26" long. I've preserved some in jars in hopes that someone working in the research of Klebsiella bacteria would be interested in studying this. I'm reading that Klebsiella is a highly mucoid bacteria and my doctor had told me that it is possible that the klebsiella are responsible for these mucous strands...any thoughts on that?

I still continue to pass some of these mucous strands, much less than at first, but still some.

I feel that the Klebsiella is the underlying issue, but I'd like your opinion...does that seem logical? I think it does, but I'm not a doctor or specialist.

I am so thankful to be regaining my health in significant ways...I hope that by sharing my experience, it may help others in some way. Yours Chris Jestus.

Chapter 99

ALL HUMANS ARE BORN PHYSICIANS

The follow-up shock comes with a rush, bringing with it multitudinous questions, some of which have no answers today. There are not, as yet, any little doors to the brain that can be opened so that a clear view of the why, when and where of the billions of its functions can be seen clearly and perfectly understood. But reality is there in the form of a frantic human being begging for understanding and begging to be understood. Uncertainty is the most painful aspect of any serious illness. It leads to wild speculation and often too much undue grief.

Edwin B. Jelks.

June was perplexed initially as to how she should respond to Chris's letter as her letter could be obviously labeled as that of a hypochondriac by contemporary medical school teaching but, then she had also studied in a different school part time. Joatmon's teachings in patient centered medicine always emphasized that all humans were born physicians by virtue of their ownership right on their systems. The role of a formally trained trouble shooter was simply to act as a facilitator in a global understanding of the nature of the human system and its afflictions. She wrote back…

Thanks Chris, for giving me the opportunity to go through your case sheet. I believe I have been able to solve your problem. I enclose a paragraph from Harrison's Principles of internal medicine below:

Klebsiella species are broadly prevalent in the environment and colonize mucosal surfaces of mammals. In healthy humans, K. pneumonia colonization rates range from 5 to 35% in the colon and from 1 to 5% in the oropharynx; the skin is usually colonized only transiently. In long-term-care facilities and hospitals, colonization occurs with K. oxytoxa as well, and carriage rates are significant among both workers and patients. Person-to-person spread is thought to be the predominant mode of acquisition.

Most of us carry a world of bacteria inside our systems where they live and die continuing their cycles of existence. Your klebsiella was one such isolated from your stool where it was living a normal life with minimum damage to your system. However the troubles which started in your system since 1988 definitely signifies the entry of a virus (something which may be likened to software programs which affect our computer systems) and it really troubled you all these days starting from an acute diarrhea, cramps, moving on to Leg aches and fibromyalgia symptoms followed by constipation which made you suffer so much.

There is nothing in your medical investigations to suggest Ankylosing spondylitis as the X-ray changes described in ankylosing spondylitis are very much different from the ones described in your imaging findings. They are non-specific changes possibly normal for your age...by the way I don't know your age accurately. Although klebsiella antigens are related to ank-spond they are hypothesized to cause so by acting as a trigger for an autoimmune reaction to the antigen by the body's cellular defense system. MUCUS is a normal constituent, which layers the bowels of all normal human beings and is passed out when stools are scanty inside the bowels. The klebsiella is unlikely to remain alive for this long inside the jars where you have stored them. Yours June.

> Over the past 600 million years the Bacteria, Archaea and microbial Eukarya have continued to evolve into brand new niches. As it happens, a few branches of Eukarya - plants and animals - grew freakishly huge bodies. They also created both new substances for bacteria to exploit, such as plant lignins, and new environments for microbes to inhabit, such as feathers and urinary tracts. Indeed, some of the richest and most interesting ecologies on Earth can be found inside the animal gut.
>
> One of the huge species, Homo sapiens, got remarkably self-important.
>
> Source: The Great Chain Of Being, Sean Nee, Nature 435, 429.

Chapter 100

A DISC FULL OF LOVE

Your love is a memory that occupies a vast space in my hard disc drive. Yet I never had a problem sharing it as an attachment every time I sent mails.

Hi! Abraham, nice poem…Do tell me more on life there now, more so because you seemed to have got over your initial teething crises. June

Dear June,

I have not got over teething crisis, I hate every day I spend in UK. Currently am doing locums in and around London-specifically now I am at hilling don hospital at

Oxbridge near Heathrow will be joining at Hastings (on the coast!!) as trust middle grade (promotion?) from 1 September u know I feel like running back to India every day.

When are you going next to Kolkata? Shall try to meet you then (planning)

Abraham

It's strange, that what you cherished forever as a dream, going to UK and working there should turn out to be a let down. Coming to Pokhara, Nepal I had a dream and it's nearly been realized in my day to day living here with the students, medicine and mountains and I am glad of my choice. I really want to meet you and have a glass of beer with you in London. I am sure we shall live to see that day soon. June

Dear June

You can come down any time u want. Why don't u in fact? If you want to give the MRCP, why don't you plan it while I am here? Or research? Well, I am ready to sponsor your trip.

Think over it.

Abraham

I just want to come there to meet you and have a nice time not give MRCP or build a career in UK. I am already happy with what I have attained in medicine (I am an internist) and my next target is to set up a community based rural practice somewhere in India (Gujrat?

Maharashtra? …suggest some place) and at the same time be attached to a rural medical college, have a house with a garden and some peace and quiet (Karnataka it will have to be I guess).

June

Chapter 101

YOURS SNEERINGLY, CHEERIO, CORNFLAKES, GRITS

Dear Joooooooon,

You just described what a general practitioner GP does here in Surrey, Yorkshire, or Devon, or Durham, or anywhere in this country…do u realize this? Think carefully I am not joking. There is opportunity in this country as a g p – they have a shortfall of 22000 G P. posts in the next 10 years. A GP earns around 100000 pounds per year and the new contract enables opting out of weekend on calls and stuff like sharing on calls so they are really never overworked. You don't need MRCP to become a GP. A GP earns more than a consultant here. I do not understand why you should forgo easy money there for the taking- u need to work only 3-4 years in UK to save approx 50-60 Lakhs in Indian money -enough for starting something inspiring in India -(in Goa perhaps!!!--I will surely join u for that)—I am not exaggerating - I saved 10 lakhs+ in 7 months - it was tough for me but its not for nothing that I have returned- after a lot of thought !!

Well its not that I am trying to hard sell something on you but I feel u will do wonderfully in the system out here where they value research, publications, audit

so very highly -(which I hate).I sense money out here and the difference with India

is amazing -- you should try to earn some out here. You deserve it actually more than some of our batch-mates who are here. Well that's all I can say - if nothing imagine what 50 lakhs can do for your security in life- you will not have to worry about your financial security for some time and what the hell where is your wanderlust gone? Don't u want to see Wales and Scotland and lake-district and Brighton (went there yesterday with Poorvi and Neil) not to speak of Switzerland and Paris and Naples and lovely Greece and Spain and canary-islands. What the hell I say!!! Where is your dammed sense of adventure??

Come baby there is gold out here. I am not drunk now. Its just I feel u will do very well here. So dammit a little trouble perhaps for some time – a hiatus, when u give PLAB but u will come flying thru. I want to see u in UK - from there we can go to USA

Hey-hey-hey, I wont take no for an answer. Abraham

Dear Abraham,

Money or comfort of security was never a priority. If it was so I would have chosen to go to UK instead of coming to Nepal. Believe me except for a few glitches like unavailability of a telephone (I have only an intercom connecting 200 people, that allows me 5 minutes of email time everyday) I don't have any qualms. I want to live life and one realization is that life keeps eluding us if we run after securities and the more money we have the more insecure we feel. Let's hope that I can live up to my reputation of being a jack of all trades (and one who doesn't need to be master of any).

June

Dear June,

All said and done I don't believe you. I don't believe u think you are good enough to compete with the best in the world, work under impossible conditions etc-I believe you are scared of fighting it out like we are doing .You have proved nothing to me or to the rest of the world - u are in a fucking backwater even worse than Kolkata, so if you have the courage then come and prove it in UK for a start. I believe u are frightened of open competition well if you have the guts come and fight and prove it-- otherwise fuck off to your rapidly crumbling make believe world of Nepali fantasy. Yours sneeringly, Abraham.

Chapter 102

Yes! I Do.

Thanks Abraham, for all the venom I never thought my dear old friend would hurl at me but it's great that we be honest with each other rather than taking recourse to the make believe. I am slightly suspicious that it may not be you because the write up lacks your trademark uppercase but who ever it may be I thank you for this opportunity to bring out the bigger truths we generally avoid/are uncomfortable to discuss verbally fearing unnecessary confrontation (something which is easier done in email rather than face to face or even over the phone).

Yes, I think you are right. I probably will not be able to carry on doing all the wonderful stuff I imagine I am doing here, especially because I will not be allowed to touch all their sophisticated equipment. I am definitely frightened of the global open competition; the very prospect of failing (which I am bound to) is sickening to say the least. Also I have a great make believe future to look forward to. However I have nothing to prove to this real world or myself. I am glad that we are finally beginning to talk and understand each other. You will always be my dear friend more so because it was your love for me that gave you the necessity to sneer at me.

Peace and keep writing

June.

Dear June

Shocked eh? But did it do what it was supposed to do? Goad you out of the hypoxic trance like security in the heights of the Nepal Himalayas? Well if it didn't its too bad, cause I think I am right.

Cheers
Ha-ha-ha

Get out of your hypoxic Himalayan stupor and realize that life is passing you by. Where is the wanderlust and daredevilry which you had? I am pissed off at your lack of sense of adventure.

Cheers

PS: Why don’t u give your Phone number to me please? Dear Joooon! For heaven’s sake, and this is the second time I am asking you, Will you marry me?!
Abraham

Yes! I would. Jooooooooooooooooooooooooooooon!

Chapter 103

THE LAST HONEYMOON WITH THE MOUNTAINS

The honeymoon was to be decided after the official marriage formalities that Abraham found too bothersome now that he couldn't wait to be alone with his beloved June. He wanted Goa again but June was firm. The mountains it would have to be and that to Macchapucchare for one last time before June packed her bags for Kolkata again. Abraham had got a nursing home attachment there and it would be convenient for him to chop-chop every day whatever appendices, hernias, and hydrocoeles he could lay his hands on. June had once fancied herself as a surgeon but once when she had taken her brother complaining of chronic pain abdomen to a very impressive professor of surgery in her medical school he immediately posted him for an appendicectomy coming morning in his nursing home. Their father, Samsara, wouldn't have any of it and said the pain was just an irritable bowel and would go away on its own. Dupki did out grow the pain once he went to college but the incident made June think that surgeons needed to be robot-like if they were to be deemed a good hand. To keep their hands in shape, they had to cut as many appendices they could lay their hands on, if not for the money it would bring in. The more the merrier, the faster the better- An intellectual surgeon was a failed surgeon.

June's Diary

26/10/01: Set out walking to Baglung bus-park with Abraham protesting vehemently (he wanted to cover the city stretch on car) and then after a 2 hour bus ride through hills and valleys finally came to Nayapul ...Verified our permits in Birethanti and took a quick breakfast before we resumed our walk.

Met my old patient of Diabetes suddenly on one of the villages in the trail...

He treated us to the guavas and oranges freshly plucked from his own garden.

However he looked quite bad, much worse than what I had seen him in the hospital .He had already developed End stage kidney failure due to diabetic nephropathy and he looked all bloated. Also what was particularly worrying was his inability to sleep at night due to the shortness of breath he had while lying down. This meant his kidneys were not excreting water properly and that was all clogging his lungs.

The guavas were tasty...and we sat relishing them near the Bhurundi river where we also spotted a white capped redstart wagging its tail. When we decided to halt for the night in Tikedhunga it was already late and Abraham immediately hit his sleeping bag but again had to arise out of politeness for two Israeli girls who had come to inspect our rooms, whether it was as bad as theirs and then we had a lot of interesting discussions on the Gaza strip and the rising US allergy among the Israelis. They liked Abraham's name as it sounded Biblical and they said a Messiah would come as a savior for all of them according to the Bible...I made them a bit uncomfortable by asking them whether they really thought it was true.

Chapter 104

REGARDLESS, WE CONTINUE...NEW OR RENEWED TRACK

June's Diary

The next day was a steep ascent to Ghorepani with tremendous views of Annapurna from Ulleri. I met an English couple coming down from Ghorepani whose guide's mother had been my patient. The lady introduced herself as a GP practicing in Cambridge and at the same time teaching communication skills to other doctors. Her husband was a dear young old man with a silver beard and a book on birds of Nepal and looked more like a bird man although he was a professor of architectural history. I wanted to talk more with them on communication skills but they were going in the opposite direction.

Whatever communication skills in med I learnt was from my mother (Who isn't a doctor).She told me what they would have expected from their own doctors and never got. She described the number of occasions she had been inadvertently offended by her own doctors. That provided for me an early start in building rapport with patients, right from my third year Med school because in effect I was avenging my mother by being good to my own patients.

The rest of the trek to Ghorepani was steadily uphill but it was made easier by Karen and her husband from California who discussed a variety of things under the sun (from surface topologies to obstetrics).We had lunch interspersed with sessions of bird watching with redstarts, minivets, orioles and kites making their forays into our imaginations if not into the nearby trees. That night in Ghorepani we had a feast of pizzas, rostis, chocolate pudding and cake but that didn't make the next day's job of climbing to Pun hill any easier although the sunrise was breathtaking. Found a few ladies photographing themselves against the background of the Annapurna baring their backs with ANNAPURNA 2003 written on them. Karen and her husband set out in the other direction towards Annapurna base camp after which they would be going to Khumbu to climb Island peak.

We soon set out for Tatopani after a quick breakfast and met Curtis and Jean Viev on the way. Curtis was a web designer but his first love was music and was soon going to bring out his first CD but before that he would be taking a month's course in Tibetan Buddhism in a monastery in Nepal. The climb down to Tatopani was tough and at the same time beautiful with fields of kodo and fapar gleaming in the afternoon sun, which was interspersed with rain that caught us on the final descent from Santosh lodge. The steep stone steps felt very slippery.

Chapter 105

ALSO SAW A BIT OF THE HIMALAYAS

`Cheshire Puss,' she began, rather timidly, as she did not at all know whether it would like the name: however, it only grinned a little wider.

`Come, it's pleased so far,' thought Alice, and she went on. `Would you tell me, please, which way I ought to go from here?' Lewis Carol

June's Diary

The next day was a quick descent to Beni as I already knew the way from last year's Muktinath visit. On the way I met a Norwegian family who were working in Nepal through generations and it felt really interesting to be able to chat all the while in fluent Nepali with people who looked European. Lunch was dashain ko maasu and kodo ko chang(a millet beer) and after reaching Beni found there were no buses since morning as it was the 10th day of Dashain.

A bus did start at 4:00 PM charging exorbitantly although we had to make do with sitting on the roof- top, which too was crowded. After some time however the roof top was gradually transformed into a place to be especially with the near full moon shining on to it a fair share of its gleaming light. It was cold but then somebody was passing around a bottle of apple brandy fresh from Marpha, Mustang.

I met Soul an Australian medical student who was a year junior to Andy who had been doing his elective with us last year and had done the Annapurna circuit in a fairly short time. Then there was this French painter whose medium was oil on water and I wondered whether it was possible to paint using only words. We lay down on the roof top of the bus watching the moon and the stars and the snow peaks dazzling at a distance while the bus moved at its gentle pace.

I felt as if I had seen the world in these few days, more so through the eyes of its people from various countries and off course I also saw a bit of the Nepal Himalayas.

Chapter 106

FLY TO KOLKATA THEN FRY

`That depends a good deal on where you want to get to,' said the Cat. `I don't much care where–' said Alice. `Then it doesn't matter which way you go,' said the Cat. `–so long as I get SOMEWHERE,' Alice added as an explanation. `Oh, you're sure to do that,' said the Cat, `if you only walk long enough.'

Lewis Carol

June couldn't manage a teaching post there but got an attachment in a clinic which was attached to a nursing home. She was still a consultant though. She didn't have any medical students to teach but then patients always needed teaching as 'docere', the Latin derivation of the word doctor suggests. June sometimes wondered if she would like it if her refrigerator mechanic came and delivered a lecture on refrigeration and maintenance to her instead of fixing the damn thing. Unfortunately physicians were far from expert mechanics and had to cover up their ignorance on system architecture and pathology by trying to share the diagnostic and management uncertainties with their patients whose systems they were supposed to repair.

Physicians were nothing but partners in trouble shooting their patients systems, part of the responsibility for which lay with the patients themselves. Like the time June used a hammer to break the ice in her deep freezer that wasn't frost free and hit one of the pipes containing some liquid gas she never knew existed. The mechanic chided her for being careless and it was obvious the whole fault was June's due to her ignorance about the mechanics of her freezer for which she had to pay a price. She remembered telling her patients in the clinic day in and day out that the anti-hypertensive drugs they were forced to gobble every day was to prevent damage to their vital organs like kidney, heart, brain, the pipelines of which were constantly being hammered with the high pressure inside them. They weren't supposed to feel any better on those drugs. It was just to take care of the silent killer hammer inside them.

Kolkata was hotter than what it used to be in her medical school days, possibly because she had grown so used to living in the Himalayas but then it also provided her with an opportunity to explore her manifold amateurish pursuits on birds, poetry, music, stars and theatre. On one of these gatherings below the stars she met Abhay Mitra, a software engineer who also shared her passion for the mountains, music and poetry. She took off on a rock-climbing trip with him to Sushunia, the place that still contained her parental memories long after their death.

She told Abraham she was going with this group who needed a doctor for first aid and he didn't seem to mind. He had himself gone on these trips in his student life as a student doctor looking after the group's first aid, taking a few classes on it as part of the rock climbing course. It was generally a good outing at low cost. Abraham was too busy in his new avatar as a foreign returned surgeon who had garnered a lot of experience there in cardiothoracic surgery.

His nursing home (well equipped from funds loaned by a wealthy businessman) soon started becoming a Mecca for those wanting quality bypass surgery in their city instead of going to other better centers in India. Abraham was soon going to become a household name in Bengal not only for his extremely amicable behavior with his patients but also exceptional robotic cutting skills along with a good skin stitch to match. The external signature of a job well done inside should be a good skin stitch he emphasized to his house surgeons. It was natural that he shouldn't have time for June, preferring to solemnly finish off the daily (or weekly?) sex routine as quietly as possibly (not allowing his face to show the slightest tinge of satisfaction).

Abhay on the other hand would spout poetry while gently caressing a rock before they began their climb in a winter chilled rock face. He could be inside her and yet sing Indian classical with a smile on his lips that indicated the immense pleasure he was deriving from it.

Chapter 107

CONVERSATIONS WITH A MOUNTAIN

The following conversations between June and Abhay lying in that rock face (after they have wriggled on top of it climbing up a chimney in its final leg) have been cooking on their portable red gas cylinder. It's been a perfect bivouac and there's no one to watch them have sex except perhaps the stars.

Abhay shows June how the moons of Jupiter look through his portable electronic telescope. "There are some people who never get to see a mountain. All they can do is simply wait. Any of these days it might just happen, flapping its wings out of the blue would appear a shining desolate golden mountain and lift them off their park side benches". June stirs the chow. "In an extremely manicured city you sometimes wish the road behind your house led to a nearby mountain. Every day on your return from office, you'd take a walk on its steep slopes. Everyday…in this extremely plain city

One longs for a very own dear mountain."

It has started raining all of a sudden and Abhay carries June down the chimney into the cave through which they had wriggled up. He bites gently into a slithery chow from June's spoon. He could recollect when it had rained last like this.

"The rain stopped for you, to open that window in the bus, for your eyes to feast on those rice fields beside the road and at a distance a shining ivory tearing through the clouds".

June had dropped her head on Abhay's knees.

"I strongly believe amidst us somewhere, resides a gigantic mountain, full of perennial snow, shining brightly in the afternoon sun, changing colors with time, morning…a pinkish white, evening…a orange hue. We go about all our lives trying to scale our individual mountains. Somewhere deep within them we exist…ourselves." Abhay decides to give June a vigorous back rub.

"Let down your hair in the wind mountain. Let it flow like a torrent through poetry swept off false moorings. Your volcanic lips let it burn each syllable and let it cool with your icy stare." June sighs and buries her head deeper into Abhay's lap.

"This mountain is an institution. One staying here too long tends to gather roots and branches and finally …one day…becomes a tree which nests innumerable eggs of countless dreams, each day we struggle to hold on to our mountains, our roots, our dreams, losing our smiles in the process. In the end the institute crumbles and leaves us

to fly on the wings of freedom" Abhay's hands shift to her hair again.

"You resemble the mountain of my dreams. Is that a running stream or your hair blowing in the wind? The day our eyes met and clouds burst into rain I had to carry a wet heart all the way to the top". June fondles his cheek.

"What are those pock marked scars? Did some one use pitons on your body? Abhay rubs his beard on her breasts.

"One day I too had longed for a climb atop your majestic tresses. Every day I had been saving a jumar or a karabiner. All is lost today in your flowing hair, a mountain stream that floods my dreams in the plains."

Chapter 108

BECOMING THE MOUNTAIN

June sits up and looks at the spotted owlet that's been watching them.

"One day, they show a mountain on the TV screen, walking towards a lonely traveler.

It grows bigger and bigger and the traveler, smaller and smaller until…the whole screen was full of the mountain. Snow covered soft powdery white and the traveler…One of the black dots on the tube…yet, nobody really ever moved from their individual places."

Abhay tells him about the first girl he had met in Shantiniketan. A beautiful slender creature in plaited braids who had walked away with a lanky goat.

"Oh! Mountain she's not for you. She's just a bolt from the blue. She's a thundercloud that will drench you from head to toe. After a time you'll notice the smiling sun point out to your cloud over another distant mountaintop." June decides to stretch out on their table top rock face. She watches the water molecules slithering down into the drop below. "For me, I am sure are waiting a few playful mountains. One has to cross them on the way to the biggest of them all. Its peak…lies invisible amidst clouds. Its surroundings…serene, silent and yet…there's this welling of love. Shining mountain…do stay put I pray. For you, I shall come definitely, one day.

Abhay tells her about his childhood. "Imperceptible and yet a gradual accumulation of hatred, anger and distrust…human relationships, they form a thundercloud of emotions from a wisp of smoke. Far down from the valley…all within seconds springs forth a gigantic mountain of misunderstandings from mole hills of misconstrued notions. June tries to catch up some sleep in Abhay's lap. "In terrible tensions sleep is only possible on listening…to a far off melody that emanates from deep within a desolate, distant mountain. There's a hint of the smiling sun and at the distant mountaintop there exists none of our dreams and wants, only sleep…deep and restful.

Abhay tells her how forlorn he used to feel after returning from his peak climbing expeditions. "On returning from the mountains, one roams the city…full of visions. Giant skyscrapers resemble Makalu…Lhotse. Vehicles ply on disgruntled streets…remind you of hurrying Sherpas on the Khumbu icefall. June tells him of the time she had spent in Lahaul-Spiti a veritable Tibetan landscape. "One day…maybe on a trip to the mountains, one might just encounter crumbled rocks and dust, on both sides of a road which extends into a never ending desert. A gradual wear and tear through the ages…plays of air wind and water. Cities buried deep within them…houses that twinkled once at night, atop dense dark mountain shapes.

They talk about city life in general. "One day just before elections you chance upon political workers painting graffiti on the mountain. Slowly and steadily it gets covered in slogans and speeches. As you go near it becomes illegible…childish scribbles. You leave and from a distance… see only a sparkling sun. Abhay tries to think how it would be living together. "Everyday I return to you, like you are home waiting, with a sweet smile and a cup of tea, the sea gulls fluttering in your eye lashes, conjure incredible patterns of white. I am sucked into them in a vortex of river rapids." June decides finally its time to pack up. "If one can't go to a mountain, it comes to him. I've been waiting for a long time now. Molehills cover my body. The Earth on it sprouts Trees. With the passage of time it seems I have become the mountain."

Chapter 109

THE POOR OLD RICH MAN

June would sometimes reluctantly accompany Abraham to the parties his clients would invite him to. On one such occasion in the 'Grande' she met Simi, Abhay's wife who had come along with her husband. She wasn't as beautiful as Abhay was handsome but then her manner of conversation immediately made June feel at home with her and she forgot about both Abhay and Abraham. Simi knew a few interesting places in Kolkata, places where she visited on weekends.

June accompanied her once and found it was an old age home where Simi had taken the responsibility of shaving the inmates once a week. She watched her paint their cheeks with her deft brushstrokes while the rest of her canvas particularly the eyes wore a weary yet peaceful expression. One of the old men started chatting up June and after a few minutes revealed he was from Nepal to which June quickly switched over to Nepali much to Simi's amusement. It was long back he said, he had left his village and he wouldn't even know the way now, it was quite well hidden from civilization. "Did you experience much poverty there?" Simi managed to butt in.

"Oh no! cried the old man. Poverty is a disease of civilization. In jungles there is no poverty. You can't call animals poor…Less powerful than us maybe, but definitely not poor."

On the drive back home Simi asked June what she thought of the poor old man's distorted economic world view especially as she had had the opportunity to view them close amidst the Maoist uprising that was very much in the news.

"I think I agree with the old man's world view." June said quietly. "But how could you? You know that place has one of the lowest GDPs on Earth!" Simi was surprised. "I think I need to collect my thoughts on this and I shall mail you an article on what I think about it." June said quietly.

Dear Simi,

Poverty was a young boy of four who met me in the streets of a busy city in India begging for money. I nearly the same age, peered at him once and then took a guilty look at my ice cream, all from the security of my parent's feathers. After that I learnt the trick of gulping ice creams without batting an eyelid while a million such children of an impoverished nation looked on, eyes gleaming with desire.

Many years have passed since then. In my medical career poverty has been a constant observer. In a country without an NHS, I have seen young people unable to afford a renal transplant dying in front of my eyes. Hospital policies in the premier institutes of India in which I have worked dictated that only those with transplant prospects could have dialysis. The rest had to go elsewhere for dialysis and not all could afford it. Some sold their property to get the money. Later, when the end came, they realized that all they had left was the satisfaction of having tried.

I left India and came to a more remote hospital in Nepal, where I learnt much about the nature of poverty.

Chapter 110

AN ONCOGENE NAMED DESIRE

Simi, Imagine a trek to a place that is completely cut off from civilization. There are lush green forests amid steep mountains, a self sustaining economy where people grow only as much as they need for their own consumption. No doubt it is a life of hardships, what with all the daily lugging of water, the work in the fields, the cutting of grass for the single cow in the family, the outdoor defecation, the abominable access to health care with the nearest centre being a two to three day trek away with the victim or patient on one's shoulder. Outsiders may want to reach out to embrace these "unfortunate people" and attempt to reform them. But then someone who was happy tilling his land for his bare necessities will realize how poor he is. He will become aware of many goodies waiting for him out there in a world full of city lights.

That is the birth of poverty, which stems from desire. Even famines have been shown to be man made, arising from the maldistribution of food by people who start hoarding grain through fear of difficult times. We continue to coin abusive terms "developing world" or "socially disadvantaged" little realizing that it was our development that resulted in the relative poverty of others.

In these remote villages there is education, which, unlike ours, teaches wisdom. It teaches us to gather and grow food and to cook it if necessary. It teaches us to live harmoniously with nature instead of plundering it.

In the villages there is a system of health care witch doctors attempt to scare diseases through various antics. Sometimes this works, but, if not, people embrace death with dignity and equanimity. If by chance people from the villages reach the world of our hospital, which lies across steep mountain passes, they are subjected to a lot of investigations that can rapidly drain their resources.

We may be proud of our cities, but are these not just concrete jungles of human misery, where the rich live in high rises and others people who couldn't make it to the top in slums? This, for us, is development, and we want the "developing world" to reach similar status.

I hope that these villages and all that they represent still exist somewhere, replete with lush green forests and inaccessible mountain passes.

Chapter 111

A GLOBAL MALIGNANCY

Thanks June, for your write up but then I feel this echoes the sentiments of people who go trekking to the developing world and remark how happy the people are without really striving to know if they would like to be happier. You can't insinuate that we NGOs leave our work of caring for the poor. I don't think you are justified in glorifying poverty unnecessarily. India needs to have the opposite of what you proffer now that it's steadily making headway into the global market. Simi

Dear Simi,

I wasn't glorifying poverty. I am also not against the need for continuing to care and relieve the squalor which abounds in the developing world. I am sure all of us who work in the so called developing world are attempting to do just that. However I wanted to see and make others see things from a slightly different perspective. It isn't the viewpoint of a traveler who remarks how happy the poor are in such and such a country. Poor are never happy in any country. My point was to try and see why and whom do we call poor. Why have people become poor? There is possibly no remote village left in this world now, disappointing as it may seem who have been left untouched by our civilization.

Our civilization in its present form has engulfed all these places, infiltrated and stifled their cultures and has become a single malignant tumor mass. Living within that mass our thoughts are easily molded to think that "The "developed" world has a duty to serve the poor and not worsen their conditions, little realizing that its the coca cola culture which comes attached with the service is what kills us. We are averse to returning to those so called miseries now because we have seen and experienced a different world of comforts but what of those people who do not(again we do not know whether they exist but as soon as we meet them we would have easily spoilt them).

William Osler was a great physician practicing 100 years back without any antibiotics or ability to thwart death (which we have been able to achieve with modern medicine?).We have definitely developed the ability to add on a few more years to our life expectancies and yet why do our miseries increase exponentially with our advancements? Where is the way forward leading us to? Another future culture where the comforts are doubled, nobody dies or becomes old? Is the way really that linear?

Going by the ravages we have inflicted in this land mass of Earth our future can only lie in Mars. Again that too will be for a chosen few. In conclusion, does it mean animals that live in the pristine wild, who are as yet undomesticated and still able to take their own life and death decisions, are much better off than us with our lap tops and idiot boxes? They don't seem to fall ill that often or even if they do we label that as a part of the ecosystem and are wary of disturbing their natural life and death cycles. Why did we have to tamper with it in case of our own species?

Chapter 112

A BAD DREAM

June continued to visit the old age home particularly the old Nepali who wouldn't allow her to lose touch of her fluency in the language although she was aware of how poor her Nepali grammar was. At times Abhay accompanied her and June suspected Simi knew of their relationship although she never let it show on her face. She seemed contented at her own work. Once Abhay said it was her inability to have children that made her so very aloof from him.

They had been married 7 years now. June assured him it couldn't be the reason rather it was Simi's natural instinct to help others rather than revel in self indulgent poetry. Abhay was offended at this last bit of spontaneity but quickly changed track to "What if…we could marry…"June swerved back with lightening speed. "Abhay, my marriage to Abraham was only necessary to experience what marriage is once. I definitely don't fancy doing that again with all the load of unnecessary hassles of divorcing in between. Also our romance shall end as soon as we marry because marriage is all about love and love is the blandest thing on Earth imaginable. "

One of the old men who June had assumed deaf suddenly broke into their conversation in between the shaving foam. "I agree with you lady, it's too much of a hassle and dead boring at that too." "I beg your pardon!" June cried out in alarm. "Don't worry I won't tell your husband. He's a nice man, did my coronary bypass recently. I am going to meet him today to show him the operative scar, which itches a little." June separated his shirt with a tremulous hand. "Let me have a look. It's a keloid. I shall give you an ointment."

"Thanks, people call me Sutra. My surname is Dhar. I even knew Samsara your father and was present at his funeral although you wouldn't have noticed at that time. "Well it was nice to know you," said June with a voice that was trying to muster all the warmth it could after recovery from the shock. Sutra continued unfazed. "I am writing a biography of a woman physician who died of HIV. Would you care to go through it some day?" June felt a very uncomfortable shudder down her spine and Abhay who had finished shaving his old man said he was ready to go if she was and both of them left quickly without exchanging another word between them.

Sutra looked up and wiped the blood from his cheek, a careless scratch of June over a thick brushstroke. "Tsk-tsk, now imagine if she's as careless with her intravenous needles.

That night June was sleeping alone as Abraham had this long OT case that had started in the evening. She felt feverish and started having unformed visual hallucinations which in her medical jargon she may have labeled photopsia. Eventually she felt someone reading out a letter. Someone who appeared like a tiny dot that was so tiny as if all the light it reflected was getting sucked into a black hole. Its voice and email was however audible and readable.

Dear June,

As a species we chanced into you a few million years ago. Some say we evolved from organisms that were present before we arrived but no one's still clear as to how they originated. All we now know is the pre-historic dinosaurs had been wiped out and we quickly learnt to use fire, invented the wheel and subsequently the computer. We still carry our evolutionary ancestors within our body either in the form of bacteria in our guts or bacterial DNA in our mitochondria. We have been recently inflicted with a dreaded virus (just as we chanced on you, once upon a time).

Chapter 113

TREES: LIKE RECEPTORS IN OUR BODY

Where we were before entering your body (like a virus) is open to speculation... maybe somewhere in Neptune? Whether it was a sexual encounter you had with her or it was the same needle which pierced into both of you, is debatable but whatever it was our ancestors had arrived and meant to stay for a long time.

I' am one of those virus species plaguing your body. I live in one of its relatively backward areas labeled the third world. However you may not appreciate our viral economics and the place where I live in Nepal may be closer to your heart. It is one part of your enchanting body, which contains one of the best walks one can have on your surface. On one such walk through the outskirts of a village on the way to Jomsom, I came across some paddy fields.

You have as many receptors in your body as we have trees on Earth. Trees...scores of them dot the landscape green, absorbing your daily quota of food from the sun. We learnt to adjust our food requirements by tampering with your receptors and cultivated a substantial part of your body into paddy fields for our survival.

Now get ready for another visual hallucination, a formed one. As I watch the lush green, fluttering in the wind against a dark cloud, I hear a distant rumble of thunder and suddenly notice a few white cattle egrets fly past in the midst of it. On the ground below I find a farmer tilling his land, along with two malnourished cows, which are barely able to shoulder the wooden plough. There are so many species of viruses & bacteria in your body, yet we spent a considerable amount of time trying to unearth newer species alien to this Earth

Definitely, there must be many of our brethren occupying other bodies as much as we are sure of the other Planets in this Universe. Recently you discovered a new species of bacteria, plaguing your stomachs. We had been using it to plough our fields for centuries. These spiral earthworms were a boon to our fields before we went into full scale mechanization and built earthworms with artificial intelligence that wreaked havoc with nature. One fine day we discovered a harvest of metallic receptors with concrete fruits, unpalatable and unfriendly. This was apart from the erratically dug out ulcers they created in your stomach antrum. It's good that you have decided to eradicate them but what will you do with us.

Chapter 114

APOPTOSIS: PERISHING GREAT CIVILIZATIONS

I am aware, we are the present scourge of mankind (in this universe) and I won't pretend any thing in our defense. Call us names. HIV or whatever code words you'd like to, sound an AIDS alert. The fact is we are doomed along with you. Off course it's not your fault, we brought it on ourselves. You might feel guilty about your casual affair with Neptune, but we were only looking for a means to survive. We landed on your virgin soil enchanted but not entrusted with a power to see the shape of things to come.

As I look up from my vantage point near these green fields I also become aware of the surrounding mountains, tall green giants preparing to see off a pristine sunset. CD4 cells you call them. Few of them have signs of habitation. A lot of trees have been chopped; the virus must leave its mark. I wonder what will happen one day when all the trees are chopped off and the mountain is fully occupied by our fellow HIVs living in concrete jungles. You know what will happen, they'll simply crumble to dust. Great civilizations, which once twinkled atop dark mountain silhouettes, will perish. You label it apoptosis; I find it a colossal waste. You feel we are invaders destroying your CD4 citadels and we are perplexed at the way things are turning out. After all we were innocent viruses looking for shelter.

With the CD4 Mountains having been squashed, the death knell has been finally dealt and not before long you'll be symptomatic. Alien beings some of us were tirelessly searching will suddenly appear in a distant night sky, huge flying dinosaurs, glowing in the dark, spitting fire. They'll land on our cities, drink all your water, uproot all trees, dig up the whole place and finish off the ozone layer till you freeze into an ice age.

Opportunistic pathogens you call them, for us they are the end. We were a peculiar race from the very beginning, thinking, and innovative, fun loving on one hand, fearful, destructive and full of hatred on the other.

All said and done, it was fun while it lasted. We pass our last days now in exile, in whatever jungles and caves spared by our brethren and the dinosaurs. Places such as this where I am standing now, on the way to Jomsom and as I stare into the firmament, Venus starts peering over the white outline of Dhaulagiri. It's a pity; most of our viral economies are tuned to measure our wealth in terms of our concrete acquisitions.

A person who lives in a cave or hut is poor by most of our standards. We simply have to acquire more and more, more TV sets, computers, washing machines, as ads selling soft drinks proudly proclaim, "Gimme more". As the sky darkens and Venus becomes increasingly dazzling I wonder why her beauty doesn't turn you on. Virgin Venus, no concrete, no dinosaurs, only lush green forests, running streams, a harmless ecosystem. Like your past. Each day I sleep with dreams of you wooing her to bed. We'll be more careful with her, it's a promise.

Very Truly Yours,
Sutra (HIV-man) or if you prefer, Death-the storyteller

Chapter 115

CREMATION-INCINERATING JUNK

As the population of the city mounted, people realized that trees were not dying at the same rate as people. They were not being born at the same rate either. It was wise of the city dwellers to devise a different method to dispel its dead. One, where the dead one's next of kin wouldn't have to watch the firewood flames by the river going up in smoke for a long time. It was time that was more important as most people continued to have less and less of it. The electric chimney came as a convenient solution. The dead bodies were brought into it in an assembly line and when your turn came all your kin could see was a sudden glow as your body caught fire and poof! That was it as the iron doors slammed shut. A few hours later your relatives would be handed your ashes in an urn.

The precious remnants of silicon molecules from your charred notebook… June had watched Samsara's body burning by the river side for a long time. There were no electric chimneys in their town then. Maya had been telling June about a swelling in Samsara's right lower limb while June was in her hostel busy preparing for her post graduate entrance exams. The exact nature of the swelling was not discernible on phone and when she inspected it on the weekend Samsara's right leg was like a log. June rang up her professor of medicine who agreed it was likely to be a femoral vein thrombosis. Samsara required urgent ICU admission for infusion of heparin molecules that June would handpick from their hospital garden. It would have prevented the thrombus spreading into his lungs, a kind of secondary prevention strategy important nevertheless, considering the life threat it would pose once it reached the lungs and blocked its circulation.

Samsara didn't want to go to the hospital he loathed all his life right from his MBBS days. He didn't want to lie there helpless on the hospital bed with his batch mates lording over him as topnotch consultants. He called over his Baul friends and asked them to sing his favorite 'Kaalia', the *Lila* of Radha-Krishna, Jivatma and Paramatma. A vivid portrayal, he believed of the interaction between the individual and the universal.

As June watched Samsara's body burning she could feel the molecules of carbon, nitrogen and oxygen ascending into the firmament. She figured they were accumulating in the clouds and would one day burst forth as rain. Samsara would burst forth on their little town once again or maybe further away from them who knew…Abraham didn't have such time for introspection about June whose body had landed up in the city chimney. He had a busy OT list the next day. Thank Con for the work-alcohol that lets us forget the past without harming our livers or our careers.

In one pocket keep a message that says: `You are just dust and ashes.' In the other pocket keep a message that says: `The world was created just for you." Both messages are, I suggest, equally true.

Richard Smith,
BMJ, 2003; 327:1430-1433

EPILOGUE

RESURRECTION: REASSEMBLING YOUR LOST PC

That was the story of Con's notebook right from creation to cremation, assimilation to incineration. It was the story of our nerd Con and his notebook body as his sole source of existential sustenance on an Earthy plane. Now that his body is no more you can see him going around like mad over hills and vales shouting yoo-hooooooo! June…Are you there? I understand if he wants to re-assimilate it he shall have to begin somewhere in the clouds. That's where most of June's molecules may still be stashed up in the upper reaches of the Earth's atmosphere. We leave it for Con to decide. S/he is not of this Earth but from a different plane altogether where PC assembling at a micro-molecular level is supposedly child's play.

ACKNOWLEDGMENTS

My partner Trinket for her conditional support and for being my balance wheel, my child for being what she is, my parents, sister and brother (both in and out of law), particularly Nami, Riker baba and luntu-shuntu for not getting to read my narratives yet. June (titles), Cecil (character narrative), Bonnie (character narrative), Curtis, Jeanne (character narrative), Trevor (character narrative), Paul (quote), Angelee (Haiku) for their valuable contribution to the narrative. Partha, Barun, Utpal, Dada, Suvo, Kaushik, Asok, Cecil, Jnana, Suvro, Basu, Laltu, Pranab,Akshay, Janette and a number of my medical students turned friends for being an inspiration.

REFERENCES

General: Longmore, M., et al. Oxford handbook of clinical medicine, 5th Ed. Oxford, Oxford university press, 2001. A fantastic read with great chapters that give us a lot to think in medicine. I have used quite a few quotes from it in this book. Unfortunately in this multi-author team write up I can't exactly pin point the authors responsible for the quotes.

The decision tree: a) Lipsitz, L.A. and Gold Berger, A.L. Loss of complexity and aging: potential applications of fractals and chaos theory to senescence. *Jl Am Med Asso*. 1992; 267:1806-9. b) Goldberger AL, Non-linear dynamics for clinicians: chaos theory, fractals and complexity at the bedside. *Lancet*, 1996:347;312-14.

Letter from a virus a) Loefler, I.J.P., microbes, chemotherapy, evolution and folly. *Lancet*, 1996, 348: 1703,04. b) Sleisinger and Fordtran, Gastro intestinal and liver disease 6th Edition, Vol2 :1132. c) Acharya, S.K. Mechanisms behind hepatitis B virus persistance: the search continues, *Indian Journal of Gastroenterology*, Sept. 1998, 17: 76. d) Ando, K., Guidetti, L.G., Corny, A., Ishikawa, T., et al. CTL access to tissue antigen is restricted in vivo *J. Immunology,* 1994, 152: 3245-53. e) Thomas, H.C., Foster, G.R. and Slumiya, M. Mutation of gene for mannose binding protein associated with chronic hepatitis B virus infection. *Lancet*, 1996; 348: 1417 – 9. f) Nagaraju, K., Naik, S.R. and Naik, S. Functional implications of presence of HBs Ag in the T cells of chronic HBV carriers. *JI Viral Hepatitis*, 1997; 4: 221 – 30.

Biswas, R. Always a medical student, Student BMJ (UK) vol 11, feb 2003, pg 41.

Chapter: Fever bugs in System Rivers. a) Biswas, R., Bhardwaj, A., Aggarwal, R., Pai, C., Krishnand, G., Sen, T., and Shetty, K.J. An Unusual Paratyphoid fever, *J Assoc Phy India*, Apr2001 Vol. 49, 477-78. b) Biswas, R., Dhakal, B., Das, R.N., and Shetty, K.J. Resolving diagnostic uncertainty in initially poorly-localizable-fevers:-a-prospective-study. *Int J Clin Pract.*(UK) 2004 Jan;58(1):26-8. c) Biswas, R., Dineshan, V., Narasimhamurthy, N.S., and Kasthuri, A.S. Integrating hospital-acquired lessons into community health practice: Optimizing antimicrobial use in Bangalore. *J Contin Educ Health Prof.* (US). 2007 Spring; 27(2): 105-10. Chapter: Cranky irritable bowel, A dysfunctional dyspeptic stomach and Setting gas purely unconditionally free. a)Lindell, G.H., Celebioglu, F. and Graffner, H.O. Non-ulcer dyspepsia in the long-term perspective. *Eur J Gastroenterol Hepatol.* 1995;7:829-33. b) Parkman, H.P. and Cohen, S. Heartburn, regurgitation, odynophagia, chest pain and dysphagia. In: Haubrich, W.S., Schaffner, F.and Berk, J.E. Eds., Bockus Gastroenterology 5th Edition, Philadelphia, WB Saunders Company, 1995; 30-40. c) Lipp, R.W., Schnedl, W.J., Hammer, H.F., Kotanko, P., Leb, G., and Krejs, G.J. Effects of postprandial walking on delayed gastric emptying and intragastric meal distribution in longstanding diabetics, *Am J Gastroenterol.* 2000 Feb. 95(2):419-24. Chapter: A Blackening lady.

An Oncogene named desire: Biswas, R. The birth of poverty, *British Medical Journal*, (UK), 2002;325:51.

Few Quotes from Sutra: a) Biswas, R., Martin, C.M., Sturmberg, J., Mukherji, K., Lee, E.W.H., Umakanth, S., and Kasturi, A.S. (2008). Social cognitive ontology and user driven healthcare, Hatzipanagos, S. and Warburton, S. (2008). (Eds.), Social Software and Developing Community Ontologies, IGI Global Publishing. (In Press). b) Biswas, R., Maniam, J., Lee, E.W.H., Umakanth, S., et al. (2008). Electronic collaboration toward social health outcomes, in (Eds.) Salmon, J. and Wilson, L. Handbook of Research on Electronic Collaboration and Organizational Synergy, Hershey, PA: IGI Global Publishing Nov 2008. c) Biswas, R., Martin, C., Sturmberg, J., Shankar, R., Umakanth, S., Shanker, S. et al. User driven health care- Answering multidimensional information needs in individual patients utilizing post EBM approaches: A conceptual model. *J Eval Clin Pract.* (UK) Oct 2008. d) Biswas, R., Maniam, J., Lee, E.W.H., Das, P.G., Umakanth, S., Dahiya, S. et al. User driven health care- Answering multidimensional information needs in individual patients utilizing post EBM approaches: An operational model. *J Eval Clin Pract.* (UK) Oct 2008. All other quotes have been referenced right after the quote.

INDEX

B

C

D

E

F

G

H

I

J

K

L

M

N

O

P

Q

R

S

U

V

W

X

Y